INTRODUCTION TO
Herpetology

second edition

INTRODUCTION TO
Herpetology

Coleman J. Goin
UNIVERSITY OF FLORIDA

Olive B. Goin

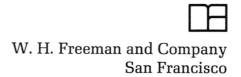

W. H. Freeman and Company
San Francisco

Library of Congress Catalog Card Number: 76-134584
International Standard Book Number: 0-7167-0686-5
Printed in the United States of America.

1 2 3 4 5 6 7 8 9

CONTENTS

PREFACE
to the Second Edition

This edition of *Introduction to Herpetology* has the same basic organization and basic philosophy as the first edition, but has a new chapter on physiological exchange with the environment, a reorganized and updated chapter on behavior, and many minor corrections and modifications.

As we had hoped, our friends have been kind to us in pointing out errors and oversights in the first edition; for this we are extremely grateful. Space and faulty memories prohibit our naming every one who has made suggestions but we must mention in particular the following: W. Leslie Burger, Roger Conant, Alice G. C. Grandison, the late Norman Hartweg, M. S. Hoogmoed, Sherman Minton, Neil D. Richmond, and Henryk Szarski. Paul Laessle helped with the art work for this edition, and Mrs. Loretta Thigpin typed the manuscript. To all of these and to the thousands of students whose use of the first edition makes this second edition possible go our sincere thanks.

As before, we hope our friends will continue to be kind enough to point out the errors of our ways. Only friendly criticism makes improvement possible.

Coleman J. Goin
July 1970   *Olive B. Goin*

PREFACE
to the First Edition

This book is planned for use as a text in a one-semester course in herpetology. It is designed for students who have had one year of college biology, but who may have had no more than one year.

Courses in herpetology usually consist of two parts—a series of lectures or discussions in which basic principles are presented, and a series of laboratory and field exercises. We believe that laboratory work should be based primarily on local faunas and on the specimens available in local institutions. These differ from region to region. Techniques useful for collecting animals in one climate may be of little use in another; the season of the year at which animals are abundant and active varies from place to place; the characters used in identifying specimens from one faunal region may not apply to those from another. Moreover, excellent field guides, keys, and local lists are available for most places where courses of herpetology are presented today. We have therefore left the organization of the laboratory and field work to the individual instructor, and have concentrated instead of the major aspects of herpetology that apply throughout the world.

In these days when so much of biology is concerned with happenings at the molecular level within the individual cell, we believe there is a definite need for the student to appreciate that these processes have biological meaning only as they help us to understand the living, functioning animal. We agree with Professor Romer that:

"It is not enough to name an animal; we want to know everything about him: what sort of a life he leads, his habits and instincts, how he gains his food and escapes enemies, his relations to other animals and his physical environment, his courtship and reproduction, care of his young, home life (if any). Some aspects of these inquiries are dignified by such names as *ecology* and *ethology;* for the most part they come broadly under the term *natural history*. Many workers who may study deeply—but narrowly—the physiological processes or anatomical structure of animals are liable to phrase this, somewhat scornfully, as '*mere* natural history.' But, on reflection, this attitude is the exact opposite of the proper one. No anatomical structure, however beautifully designed, no physiological or biochemical process, however interesting to the technical worker, is of importance except insofar as it contributes to the survival and welfare of the animal. The study of the functioning of an animal in nature—to put it crudely, how he goes about his business of being an animal—is in many regards the highest possible level of biological investigation." *(The Vertebrate Story,* A. S. Romer, Univ. Chicago Press, Copyright 1959 by the University of Chicago.)

In preparing this volume, our task has been twofold: we have had first to decide on the basic organization that we felt a text in this field should have, and second to synthesize a mountain of original literature into a volume of modest size.

In the former task, we have worked under the firm conviction that the proper approach was to discuss basic biological principles as exemplified by amphibians and reptiles. This we have tried to do. In the latter task, we have of necessity shown a great deal of personal bias in deciding just what should and what should not be included. Because of the wealth of original literature, there are many fascinating facts of herpetology that we have had to leave unmentioned. We know that some of our professional colleagues will feel that, like good Anglicans, "we have left out those things which we ought to have put in, and we have put in those things which we ought to have left out."

In the first chapter we indicate the position of the amphibians and reptiles in the animal kingdom, discuss briefly certain basic principles of classification, and give a resume of the rise of herpetology as a science. The next four chapters deal with the structure and evolutionary history of the two classes. Chapters 6 through 11 are concerned with natural history and with the mechanisms of speciation and geographic distribution. The last six chapters give a summary of the living amphibians and reptiles to the family or subfamily level, with notes on life history and geographic distribution.

Since herpetology is a worldwide subject, we have tried, in our choice of examples and illustrative material, to strike a balance between native and exotic forms.

The references given at the end of each chapter are intended simply as suggestions to the interested student who may wish to pursue a particular topic further than is possible in an introductory text. Most are compendiums. Occasionally, we have included original papers that are of exceptional interest and importance and that give information not generally available elsewhere.

It is a pleasure to acknowledge the help we have received from so many people in the preparation of this text. We are indebted to M. Graham Netting for reading Chapters 8, 9, 11, 13, and 17, and to Archie Carr for reading Chapters 8, 9, and 14. We wish especially to thank Kenneth W. Cooper, not only for the critical reading of Chapter 10, but also for the continuing interest he has shown in this work, and for the many pertinent references he has sent us. Henryk Szarski read Chapter 4 and Carl Gans read portions of an earlier draft of the manuscript. Walter Auffenberg assisted us materially in dealing with the classification of the snakes and crocodilians.

Mr. and Mrs. J. C. Battersby, Charles M. Bogert, Robert F. Inger, and Alfred S. Romer all responded most kindly to appeals for special information and material.

Not least have been the intangible benefits we have received from discussions with these and other colleagues, among whom we should like to mention in particular Doris M. Cochran and Ivor Griffiths.

Except where otherwise indicated, the excellent photographs were made by Isabelle Hunt Conant, many of them especially for this volume. The line drawings are from the gifted pen of Evan Gillespie, many of them from sketches made originally by Ester Coogle.

To James E. Bohlke, Alice G. C. Grandison, W. S. Pitt, Oswaldo Reig, and Charles K. Weichert we are indebted for special illustrative material.

We wish to thank Dean George T. Harrell for the use of a dictaphone during the preparation of the first draft, and Mrs. Sue P. Johnson for her careful typing of the final draft.

Since everything we have ever done has had imperfections, we feel sure that this book will have its share. We would like to request that our friends be kind enough to point out to us our errors, both of omission and of commission, so that in the future we may mend our ways.

March 1962

Coleman J. Goin
Olive B. Goin

INTRODUCTION TO
Herpetology

1

INTRODUCTION

The science of biology has become enormously complex. No longer is it possible for the work of one person to encompass all of its ramifications. To keep up with advances biologists have been increasingly obliged to limit themselves to various subsciences, such as anatomy, genetics, embryology, or ecology. These subsciences can be visualized as extending vertically through the parent science. But this is not the only way biology can be approached; it can be subdivided according to the kinds of organisms studied. These subdivisions include such disciplines as ornithology, entomology, and herpetology. They extend across and interweave with the primary divisions mentioned above.

Biology might thus be visualized as a vast tapestry with the threads of the warp formed by the many subsciences and those of the woof formed by the groups of organisms under study. Whether one wishes to follow a thread of the warp and study something like the anatomy of one or more structures in many different kinds of animals (comparative anatomy), or to follow a thread of the woof and study the anatomy, behavior, and distribution of a particular group of organisms (such as the snakes), is a matter of personal inclination. When we study herpetology (*"herpeton"* = crawling thing, *"logos"* = reason or knowledge) we follow those threads of the woof that are made of the amphibians and the reptiles. Any aspect of the biology of these animals is legitimately a part of the subject matter of herpetology.

ZOOLOGICAL POSITION
OF AMPHIBIANS AND REPTILES

We cannot profitably study any of the broader aspects of zoology until we know just what animals we are dealing with and where they fit in the whole pattern. No one knows exactly how many different kinds of animals there are, but one careful estimate gives 1,120,000 and this is within reason. To bring order and meaning into this bewildering array of forms, man has found it necessary to classify and divide them into groups and categories. There are a number of different schemes of classification that we might adopt. We could, for example, divide animals according to their habitat, such as forest, desert, lake, or ocean. Some studies require this kind of classification, but the standard, universally accepted classification of today, the one we mean when we speak of animal classification, is based on the degree of relationship through descent from a common ancestor. Closeness of relationship is usually judged by similarity of morphological characters.

The animal kingdom is divided into a number of large groups called phyla, (e.g., phylum Mollusca: shellfishes, like the oysters, clams, snails, and squids; phyum Arthropoda: joint-footed animals such as insects, spiders, centipedes, and lobsters). The phylum Chordata comprises animals that at some stage in their life history have pharyngeal pouches, a hollow, dorsal nerve cord, and a notochord (a stiffening rod running along the back). The phylum Chordata is divided into several small subphyla and one large one, the subphylum Vertebrata, to which belong the animals most familiar to us— the fishes, amphibians, reptiles, birds, and mammals. Vertebrates are animals in which the notochord, though still present in the early embryonic stages, has been largely replaced in the adults by a jointed vertebral column composed of a number of separate structures, the vertebrae. The anterior end of the nerve cord is expanded to form a brain which is enclosed in a protective box, the cranium.

Members of the subphylum Vertebrata are divided into a number of classes. Included are several classes of fishlike vertebrates plus the classes Amphibia, Reptilia, Aves, and Mammalia. These comprise about 38,000 kinds of living animals:

Fishlike vertebrates	17,000 ± species
Amphibians	2,400 ± species
Reptiles	6,000 ± species
Birds	8,600 ± species
Mammals	3,500 ± species

Amphibia

Amphibians are vertebrate animals whose body temperature is dependent upon the external environment; that is, they are ectothermal. They have soft glandular skins that are for the most part without scales; they lack the paired fins of the fishes and instead most have limbs with digits as do the higher animal forms—the reptiles, birds, and mammals. These four classes of animals with limbs are sometimes linked in the superclass Tetrapoda (four-footed). The amphibian egg lacks a shell and, to prevent the developing embryo from desiccating, it must be laid in water or in humid surroundings.

Ancestral amphibians were derived from primitive fish. They were the first vertebrates to move onto land and they gave rise to all the other terrestrial vertebrates: the reptiles, birds, and mammals. Two hundred fifty million years ago they were a prominent element in the world's fauna; today they are a decadent stock with only about two thousand living members. These are divided into four orders:

Order Gymnophiona (Caecilians)	158 ± species
Order Trachystomata (Sirens)	3 species
Order Caudata (Salamanders)	300 ± species
Order Anura (Frogs and Toads)	1,900 ± species

Reptilia

Reptiles, like amphibians, are ectothermal vertebrates that do not have paired fins. They differ from amphibians, however, in that they all have scales. The reptilian egg usually has either a parchmentlike or calcareous shell and is laid on land. In addition to a yolk sac, it has three extra-embryonic membranes—the amnion, chorion, and allantois—that are not present in amphibian and fish eggs. Since these membranes are also present in birds and mammals, these three classes are sometimes called amniotes to distinguish them from the anamniotes, the fishes and amphibians.

Reptiles are descendants of the early amphibians and were the dominant animals of the earth during the Mesozoic era; they gave rise to the two classes of vertebrate animals that have internal temperature control, the endothermal birds and mammals. Reptiles have lost the dominant position they held during the Mesozoic, although they are still much more numerous than amphibians. Living reptiles are divided into the following orders:

Order Testudinata (Turtles)	320 ± species
Order Rhynchocephalia (Tuatara)	1 species
Order Squamata	
Suborder Lacertilia[1] (Lizards)	3,000 ± species
Suborder Amphisbaenia (Ringed Lizards)	140 ± species
Suborder Serpentes (Snakes)	2,700 ± species
Order Crocodilia (Alligators and Crocodiles)	21 species

---

[1] The Greek forms, Sauria and Ophidia, are often used for the suborders of lizards and snakes, respectively, but we prefer the Latin forms, Lacertilia and Serpentes, because the ordinal name, Squamata, is also derived from Latin.

Amphibians and reptiles have traditionally been studied together. This is partly for historical reasons—at one time the differences between the two groups were not recognized as being important enough to justify their placement in separate classes. It is also partly a matter of convenience—amphibians are a small group and the methods of collecting and preserving them are similar to those used for reptiles. To avoid repeating the rather cumbersome phrase "amphibians and reptiles" we will hereafter use "herptiles" as a general term for the members of these two classes.

With our knowledge of biology becoming more and more detailed and the literature more and more voluminous, it is difficult for a worker even in the restricted field of herpetology to keep abreast of current developments. Some modern herpetologists restrict themselves almost entirely to the study of amphibians, or of reptiles, or perhaps even of a single order.

SYSTEMATICS AND TAXONOMY

The task of naming, describing, and classifying amphibians and reptiles has necessarily had priority, and even at the present time much of the literature of herpetology is concerned with such studies. Here herpetology interweaves with taxonomy and systematics. These two terms are often used interchangeably, but it seems better to restrict "taxonomy" to the frequently very complicated task of assigning names to groups of animals, and "systematics" to the formulation of a classification that will be descriptive of the relation of the animals to one another. The two do overlap, of course. Current taxonomic practice requires that the taxonomist describe the form he is naming and that he include in the description an indication of the relationships of the animal. The systematist frequently finds that he must name one or more new forms or resolve a nomenclatural problem before he can discuss intelligibly the relationships of the animals he is classifying.

Our current system of classification comprises a series of categories, each less inclusive than the preceding one. Phyla are divided into classes, classes into orders, orders into families, families into genera (sing. genus) and genera into species (sing. species). A Leopard Frog is thus classified:

Phylum Chordata
Class Amphibia
Order Anura
Family Ranidae
Genus *Rana*
Species *Rana pipiens*

These are the standard categories, but as classification has grown more precise we have come to recognize the need for additional groupings. So we sometimes have superclasses,[1] infraorders, subfamilies, and so forth. If a genus includes a large number of rather diverse species that fall into natural groups, it may be divided into several subgenera. If it includes only a few similar species, subgenera are not necessary. Every animal is a member of a species, a genus, a family, an order, a class, and a phylum; it may or may not belong to a supergenus or suborder.

The basic systematic unit is the species. Species are generally what we have in mind when we talk about "kinds" of organisms. Yet simple as this may seem, probably no concept in all biology has been more argued about than the species concept. Biologists do not even agree whether a species has objective reality or is subjective, a convenient pigeonhole devised by man to simplify the handling of data. Most of them do agree, though, that the species is somehow different from the other categories.

There have been innumerable attempts to define "species," but none of them are completely satisfactory for all organisms. Considering the multiplicity of animal and plant forms, and the diverse ways they have of reproducing, this is not surprising. Any definition that covered all would either be so vague as to be meaningless or would expand into an essay or even a book and would thus no longer be a definition. Fortunately, we are here dealing only with amphibians and reptiles, animals that are multicellular, usually bisexual, and not self-fertilizing. For such animals it is possible to frame a working definition: *A species is a population of animals that freely interbreed in nature (or would do so if brought together) to produce fertile offspring.* The concept of a particular species may indeed be a creation of the mind of man, but the population on which it is based is real. The conscientious systematist tries to frame and express his concept so that it will conform as closely as possible to that objective reality.

When a population is spread over a geographic area comprising diverse environmental conditions, the members of the species in one part of the range frequently differ slightly from those in another part. These different parts of the population are geographic races or subspecies. They are not reproductively isolated, since in those places where two races meet they do interbreed. The population in such an area of intergradation is intermediate in character between the two races.

The criterion of whether two forms can and do interbreed is basic to the modern species concept, but unfortunately it is often very difficult to apply in practice. Except for frogs, which proclaim their intentions to high heaven, we are seldom able to observe breeding in the field. Nor does it always help to bring the animals into the laboratory. Sometimes two forms that do not

---

[1]The prefixes ("super" = above, "infra" and "sub" = below) indicate where the category falls in relation to the standard ones given above. Thus a superclass is a grouping of classes within a subphylum or phylum.

ordinarily cross in nature will do so under artificial conditions; other animals refuse to mate at all in captivity.

Consider the plight of a biologist who has received two collections of reptiles, each from a different locality in Africa. In one is a series of snakes that are all much alike. In the second collection is a series of snakes that are slightly smaller, have yellow stripes on a black background instead of tannish stripes on a dark brown background, and have an average of 150 ventral scales rather than 170 as do the snakes in the first series. Do these two forms interbreed in nature? Are they two species or two geographic races of one species? Short of organizing an expedition to Africa and hunting for an area of intergradation between the two, how is our biologist to find out? Must he refrain from publishing on these collections and describing and naming the snakes until he finds out? Fortunately, no; otherwise we should never get on with the business of naming and classifying the animals of the world. We know from experience that, in the amphibians and reptiles at least, two forms that are reproductively isolated usually differ in morphological characters and that *in general* the morphological differences between two species are greater than the ones between two subspecies. Here the experience and judgment of the biologist come into play. He simply decides whether or not the two forms are enough alike to presume interbreeding, and then describes them as subspecies or species accordingly. It should be stressed, though, that the degree of morphological difference is an indication of, but does not determine, the degree of relationship. Sometimes two very similar forms are reproductively isolated and hence are distinct species.

Morphological differences alone do not make a species, but they *are* correlated with differences in physiology, cytology, behavior, and ecological requirements that result in reproductive isolation. We use morphological characters most often because they are the ones we can see, measure, and compare in preserved material.

Furthermore, we must always remember that species are products of evolution and that evolution is gradual. It would be surprising indeed if we did not sometimes find two related and adjacent populations that are in the process of becoming species, that have not quite achieved complete reproductive isolation from each other. Here again the systematist must use his own judgment in deciding whether or not the degree of isolation is sufficient to warrant calling the two forms species. The decisions of the systematist are by no means final. Collection of further data often shows that two forms that were described as separate species are really geographic races of one species, and sometimes that what was called a subspecies is a full species. Fortunately, under our current system of nomenclature, it is easy to make such shifts in rank without upsetting the whole scheme of classification.

The systematist must also distinguish between morphological similarities due to relationship and those that have developed through parallel or convergent evolution in forms that are quite distantly related. When two separate lines of organisms become adapted to the same basic type of environment in different regions of the world, their members may come to resemble each other closely. This has happened frequently among the herptiles. Several montane species of Asiatic land salamanders of the family Hynobiidae have reduced lungs and one genus, *Onychodactylus*, has lost the lungs entirely, a loss that is characteristic of the distantly related mountain salamanders of the family Plethodontidae. Some Old World tree frogs of the genus *Rhacophorus* (family Rhacophoridae) look so much like some South American tree frogs of the genus *Phyllomedusa* (family Hylidae) that it takes close inspection to distinguish between them. Among the reptiles, the Moloch *(Moloch horridus)* of Australia, a lizard of the family Agamidae, has an appearance very similar to that of the horned lizards *(Phrynosoma)* of western North America, members of the family Iguanidae. Both live in arid regions and feed almost exclusively on ants. There is, of course, the classic case of the Sidewinder, *Crotalus cerastes,* a rattlesnake of the deserts of western United States, that resembles so closely the Sand Viper *(Cerastes)* of the deserts of northern Africa and southwestern Asia that its specific name was given in recognition of this similarity. The likeness is further emphasized by their similar sidewinding locomotion.

NOMENCLATURE

Our system of nomenclature is based on the one first universally applied to all animals by Linnaeus in the tenth edition of his *Systema Naturae,* published in 1758. Linnaeus gave each species known to him a name consisting of two parts, the name of the genus to which it belonged, plus a specific epithet, the trivial name (e.g., *Rana esculenta*). Over the years other biologists adopted the Linnaean system of designating species. But as more and more new species were found and named, and more and more papers were published using these names, confusion inevitably crept in. Sometimes a man, either deliberately or through ignorance, gave a name to a species that had already been named something else, thereby creating a *synonym*. Sometimes two men, working on different groups, happened to give the same name to entirely different forms. Such a name is called a *homonym*. But if we, as biologists, are to understand one another, we must be sure that each species has only one name and that each name applies to only one species.

The need for a set of rules for taxonomists to follow in proposing new names, or in deciding which of several names already in use should be accepted, soon became acute. A number of codes were drawn up in different countries and for different groups. The English followed one code, the French another, the Germans still a third; ornithologists had a code of their own, and so did paleontologists. Finally, in 1895, the Third International Zoological Congress appointed a Committee which drew up the *International Rules of Zoological Nomenclature,* commonly known as the Code. This Code was accepted by the Fifth International Zoological Congress in 1901 and is now universally followed. It has been revised from time to time, most recently in 1961. A permanent International Commission of Zoological Nomenclature serves as a "supreme court" to resolve the knotty problems of interpretation that seem to be inevitable under any code of laws.

Most of the rules and recommendations of the Code deal with technicalities of interest only to professional taxonomists. A few of them should be familiar to every zoologist, if only because an understanding of them will clarify his reading and facilitate his writing on zoological subjects.

No genus may have the same name as any other genus of animals, but the same trivial name may be used in different genera (though not more than once within a single genus). The specific name of an animal consists of the generic name plus the trivial name. Thus the specific name of the Leopard Frog classified in the preceding section is not *pipiens* but *Rana pipiens.* When a specific name is written, the generic name always begins with a capital letter, the trivial name with a small letter, and both are italicized, indicated in a manuscript by underlining the name. If a species is found to comprise two or more geographic races, each is given a third name which follows the trivial name and like it is never capitalized. The third name of the subspecies that most nearly represents the material on which the trivial name of the species was based is always the same as the trivial name. Thus the race of the Leopard Frog in eastern United States is *Rana pipiens pipiens* while the western race is *R. p. brachycephala.*[1] Sometimes a scientific name is followed by the name of the man who first proposed it, e.g., *R. p. pipiens* Schreber. The name of a higher category (family, order, class, etc.) begins with a capital letter but is not italicized in print. Family names always end in "-idae" (Ranidae) and subfamily names in "-inae" (Raninae).

DEVELOPMENT OF HERPETOLOGY

Among the early writers on natural history, Aristotle and Pliny stand out. Aristotle (384–322 B.C.) has been called the originator of biological classification, not because his system bears much resemblance to the one we use today, but because he first felt the need to establish categories based on characteristics, particularly anatomical ones, of the animals. He says "Four-

---

[1]Note that when the full name has been written out once in a paragraph, we may, if we use the name again in the same paragraph, abbreviate the generic name and, when a trinomial is used, the trivial name.

footed beasts that produce their young alive have hair; four-footed beasts that lay eggs have scales." Compare this with the Biblical division of animals as "clean" or "unclean."

The Roman Pliny (A.D. 23–70) wrote a *Naturalis Historiae* in which he listed all the animals of which he knew or had heard. Pliny was not a systematist and there is little order to his arrangement. He was also very credulous. Many of his animals are fabulous and many of the tales he tells of real animals are equally fabulous. But his accounts are lively and, where he deals with animals he could study himself, they contain some accurate information.

The works of these two men were handed down in manuscript form during the Middle Ages, and were among the first books made widely available after the invention of printing in the mid-fifteenth century. Pliny's "Natural History" was published in 1469 and some of the works of Aristotle, including his "On the History of Animals," were printed in the period 1495-1498.

The reawakened interest in the world of nature that came with the Renaissance, and the enormous widening of the horizons of that world following the voyages of the explorers, resulted in the discovery of many more kinds of animals. The works of Pliny and Aristotle no longer sufficed. Several new compendia appeared, largely based on the classification of Aristotle and the style of Pliny. Notable is the *Historia Animalium,* published between 1559 and 1583 by Conradus von Gesner, a talented Swiss naturalist. Gesner's greatest contribution was the introduction of scientific illustrations. His woodcuts of a frog, a toad, a ringed snake, and a viper are probably the first published figures of amphibians and reptiles.

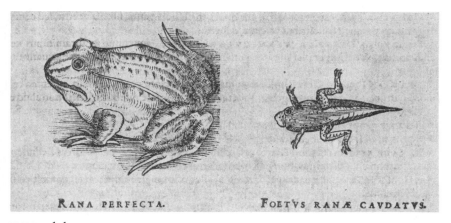

RANA PERFECTA.   FOETVS RANÆ CAVDATVS.

FIGURE 1-1

"Rana perfecta" as illustrated by Gesner in his *Historia Animalium*, published in 1586, one of the earliest herpetological illustrations. The name is not a specific name in the Linnaean sense, but means the adult or "perfect" form of the frog. [Courtesy of the United States National Museum.]

The first English language book on natural history is by the Reverend Edward Topsell who, in 1608, published *Historie of Foure-footed Beasts, describing the true and lively figures of every Beast. . . . collected out of all the volumes of C. Gesner and other writers.* The next year Topsell brought out his *Historie of Serpents or the Seconde Book of Living creatures.* Topsell's conception of serpents included practically everything that creeps and crawls and hence involved spiders, scorpions, and many fabulous creatures, in addition to the amphibians and reptiles.

The first truly critical and systematic approach to herpetology was made by John Ray, an Englishman, son of a village blacksmith. He was primarily a botanist but his *Synopsis Methodica Animalium Quadrupedum et Serpentini Generis,* published in 1693, is concerned with the herptiles. Ray was the first to group these animals together for the reason that their hearts have a single ventricle in contrast to the two-chambered ventricle of the birds and mammals. He further distinguished between the harmless and poisonous snakes by the character of their teeth. His work was excellent for the time.

It is unfortunate that Linnaeus (1707–1788) did not adopt Ray's classification. He named the herptiles he knew and gave them a place in his classification, but he had little interest in or liking for them, as is shown by his summary in *Systema Naturae*:

> These foul and loathsome animals are distinguished by a heart with a single ventricle and a single auricle, doubtful lungs and a double penis.
> Most amphibia are abhorrent because of their cold body, pale colour, cartilaginous skeleton, filthy skin, fierce aspect, calculating eye, offensive smell, harsh voice, squalid habitation, and terrible venom; and so their Creator has not exerted his powers (to make) many of them.

The classification of Linnaeus is faulty and inadequate, but the principles he laid down and the system of nomenclature he established allowed other workers to place systematic herpetology on a firm foundation.

In post-Linnaean times, herpetology has made rapid advances. We shall here mention only a few of the more important early workers. Among these is André Marie Constant Duméril, a French naturalist and herpetologist. With his colleague, Gabriel Bibron, Duméril wrote an important ten-volume work, *Erpétologie Générale,* which was published from 1835 to 1854. Some of these volumes were co-authored by Auguste Henri André Duméril.

Albert C. D. G. Günther, who served as Keeper of Reptiles in the British Museum from 1856 to 1895, gave tremendous impetus to herpetology when he published his series of catalogues of amphibians and reptiles in the British Museum. He also did spadework for neotropical herpetology in his volume on amphibians and reptiles in the series *Biologia Centrale Americana.* George Albert Boulenger, who worked in the British Museum from

1882 to 1920, laid the groundwork for the worldwide study of herpetology with a nine-volume catalogue of the batrachians (amphibians) and reptiles in the British Museum.

FIGURE 1-2
George Albert Boulenger. [Courtesy of the British Museum (Natural History).]

   In the New World, Edward Drinker Cope (1840–1897) was publishing paper after paper on fundamental herpetology. Cope not only made major contributions in his long series of descriptive papers on amphibians and reptiles, but also, through his great knowledge of comparative anatomy and paleontology, provided a foundation for much of our modern classification of these animals. Two of his most important works, *The Batrachia of North America* (1889), and *The Crocodilians, Lizards and Snakes of North America* (1900), are still fundamental books in the library of anyone concerned with North American herpetology. The latter book was published posthumously.

   The work of Leonard Stejneger (1851–1943) is perhaps not appreciated as much as it should be by many younger North American herpetologists. From the time he was appointed Curator of Reptiles in the United States National Museum in 1889, until his death in 1943, he published voluminously on the herpetology not only of America but of the world. Among his important books are the *Herpetology of Japan* and *The Poisonous Reptiles of North America*. But perhaps his major contribution was in the work he undertook with his younger friend and associate, Thomas Barbour, Director of the Museum of Comparative Zoology. In 1917 the two brought out the first edition of the *Check List of North American Amphibians and Reptiles,*

FIGURE 1-3
Edward Drinker Cope. [Courtesy of
the Academy of Natural Sciences of
Philadelphia.]

a volume that went through five successive editions under their joint author-
ship. It is hard indeed for many young herpetologists to appreciate fully the
difficulties faced by researchers in the days before a checklist and handbooks
for every group were readily available. Stejneger and Barbour, by synthesiz-
ing North American herpetology into a readily digestible form, had an
incalculable effect on stimulating research in the field.

Finally, we mention the name of Raymond L. Ditmars, late Curator of
Reptiles at the New York Zoological Society. At a time when the publication
of popular books was frowned upon as unworthy of a true scientist, Ditmars
persisted and turned out a series of popular volumes on reptiles, particularly
snakes, that awakened many a youngster to an interest in herpetology. It was
only after this interest was developed that these young workers could begin
to digest the more fundamental work of such men as Blanchard, Dunn,
Noble, Schmidt, and Van Denburgh. Ditmars was thus performing one of
the most important functions of a scientist; one that is all too often over-
looked. A scientist has the responsibility of interpreting his specialty to the
general public and of answering the questions of amateurs and interested
laymen. Since Ditmars' time, several other herpetologists have recognized
this responsibility and have written excellent popular books on the herptiles.

These books are helping to diminish the widespread, long-standing, and
unwarranted public antipathy toward the herptiles. Nevertheless, too many
people still share the opinion of Linnaeus and wonder why anyone should
waste his time on such unpleasant creatures. One answer is that a thorough
knowledge of the herptiles will elucidate most of the biologic principles that

apply throughout the animal kingdom and may provide a clue to some of the unanswered problems of basic biology. But in the last analysis, the herpetologist studies reptiles and amphibians because, in contrast to Linnaeus, he likes them and finds them interesting. *De gustibus non disputandum.*

READINGS AND REFERENCES

Carr, A. *The Reptiles.* New York: Life Nature Library, 1963.

International Commission on Zoological Nomenclature. *International Code of Zoological Nomenclature.* London: International Trust for Zoological Nomenclature, 1961.

Mayr, Ernst. *Principles of Systematic Zoology.* New York: McGraw-Hill, 1969. (A stimulating discussion of the modern approach to problems of systematics and taxonomy.)

Nordenskiöld, Erik. *History of Biology.* New York: Tudor, 1928. (Perhaps the best single book on the history of biology; excellent accounts of the lives and works of great biologists, and the relationships of each to the historical events and intellectual climate of his time.)

Peters, J. A. *Dictionary of Herpetology.* New York and London: Hafner, 1964.

Romer, A. S. *The Vertebrate Story.* Chicago: The University of Chicago Press, 1959. (A very readable, well-illustrated account of the evolution and natural history of the vertebrates.)

Smith, M. A. *The British Amphibians and Reptiles.* Rev. ed. London: Collins, 1954.

Young, J. Z. *The Life of Vertebrates.* New York and Oxford: Oxford University Press, 1962.

2

STRUCTURE
OF AMPHIBIA

If amphibians had not made the transition from an aquatic to a terrestrial existence the evolution of the higher vertebrates could not have taken place. The cause and course of this major shift of habitat have been subjects of much speculation. The individual modern amphibian accomplishes the same transition in weeks or months, instead of thousands of years, and because we are so familiar with the change in life style we accept it as one of the commonplaces of nature. However, for neither the individual nor the group is the transition complete; most adult amphibians are bound to humid environments and those that lay eggs must do so either in water or in moist surroundings. Their transitional position imposes structural and functional complications. Some of their characteristics are adaptations to life in the water, others to life on land. Some reflect the necessities of an individual life cycle that involves a change from an aquatic larva to a terrestrial adult that nevertheless must return to the water to breed. The larva, moreover, does not resemble the fish ancestor of the amphibians, and the steps by which it metamorphoses into an adult are not the same as the steps by which the fish evolved into an amphibian. The tadpole, for instance, does not start with fins that change into legs.

The amphibians have no unique structures, such as the feathers of birds or the mammary glands of mammals, that set them off from the other groups of tetrapods. There is no structure of which we can say, "This is found in all amphibians and only in amphibians." Even the characteristic life history which gives the group its name (*"amphibios"* = double life) does not hold for all members, since some amphibians lay their eggs or bear their young on land and have no aquatic larval stage, yet no one doubts that they are true amphibians. Therefore, any definition of the class Amphibia must be based on a combination of characteristics rather than on definitive, unique ones.

DEFINITION OF AMPHIBIA

Most members of the class Amphibia—frogs, salamanders, sirens, and caecilians—are small animals that have smooth, moist skins without scales. They lay their eggs in water or in moist surroundings. The egg is covered by several gelatinous envelopes rather than a shell and it usually hatches into a larva. The larva differs structurally from the adult, and metamorphoses (that is, changes abruptly rather than gradually) into the adult body form.

The larvae have gills which in some species may be enclosed within a gill chamber. In the adults, respiration takes place either through lungs, through gills, through the skin or mucous membrane lining the mouth and pharynx, or by means of some combination of these. The lungs are simple in structure and usually appear before metamorphosis.

Amphibians are ectothermic. Multicellular mucous and poison glands occur in the skin. There are no true nails or claws, although in some forms horny epidermal structures may be present at the tips of the toes. The heart is three-chambered. The skull is flattened and composed of fewer bones than is the fish skull; it articulates with the vertebral column by two occipital condyles (a condition characteristic of mammals but not of reptiles or birds). As in the fish, there are only ten pairs of cranial nerves, rather than the twelve pairs present in the higher tetrapods.

ANATOMY OF AMPHIBIA

Detailed anatomical descriptions will not be presented in this book. We are here concerned chiefly with those structural advances from the condition in fishes which, in combination, make an animal an amphibian.

Integument

We are accustomed to think of the evolution from the aquatic fish to the more or less terrestrial amphibian as involving mainly the shift from aquatic to aerial respiration and the transformation of fins into legs. Actually, the process called for basic structural changes throughout the body. Some of the most important of these took place in the integument, since the skin is, after all, the part of the animal directly in contact with the external environment.

The amphibian skin, like that of all vertebrates, consists of two parts: an outer layer (the epidermis) and an inner layer (the dermis).

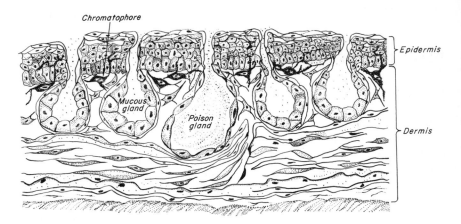

FIGURE 2-1
Cross section through the skin of *Rana pipiens*.

Epidermis.   The epidermis comprises several layers of cells, including a stratum corneum, which is an outer layer of horny, dead cells that protects the deeper living cells and helps prevent moisture loss. This is an adaptation to life on land; it is not present in fish and is lacking in only a few aquatic salamanders among the amphibians. During ecdysis (the shedding of the skin) it is the stratum corneum that is sloughed. In some forms (e.g., the Leopard Frog, *Rana pipiens*) the skin comes off in bits and pieces. In others e.g., the toads, *Bufo*) it is pulled off as a whole. We never find a shed toad skin as we do a snake skin because the toad eats it immediately. The epidermis consists of two or more cell layers before the larva hatches.

Dermis.   The dermis is relatively thin. It is composed of two layers, an outer, loosely organized stratum spongiosum and an inner, more compact stratum compactum. Since it serves as a respiratory organ, it is usually well supplied with blood vessels. Pigment in the amphibians is found in special cells called chromatophores, which are usually located in the uppermost layer of the dermis or between the dermis and the epidermis.

Glands.   Epidermal glands may be either unicellular or multicellular. The unicellular glands that are so common in fishes have almost disappeared in amphibians. Patches of them develop in the head region of the embryo and secrete an enzyme that digests the gelatinous envelopes of the egg and so aids in hatching. Unicellular glands known as the glands of Leydig are present in the epidermis of some larval salamanders; their function is unknown. On the other hand, multicellular mucous and poison glands are numerous and well developed in amphibians. The mucous secretion helps keep the skin moist and provides a medium for the exchange of gases, which is very important for animals depending wholly or partly on integumentary respiration.

Poison glands are well developed in many different species, especially in the more terrestrial anurans. The so-called warts of toads and the parotoid glands on the back of the neck are actually masses of poison glands. *Hyla vasta,* a giant tree frog found in Haiti, may give off a poisonous secretion so strong that the unwary collector who picks one up with bare hands may suffer inflammation of the skin. Many herpetologists have found all of the frogs in their collecting bags dead at the end of the day except for one species. Apparently the poison of one species may be lethal to members of other species but not to individuals of its own kind. Of course the story of getting warts from handling toads is fictitious.

Some specially modified integumentary glands are present in frogs and toads. Many of the tree frogs have glandular discs at the tips of the fingers and toes which apparently aid in climbing, and the glandular thumb pads of breeding male anurans help them to clasp the females. Male salamanders may have hedonic (pleasure-giving) glands either under the chin (mental glands), on the face, or on the tail. The secretions of these glands stimulate the females during courtship.

Scales.   As a general rule, the dermal scales that are characteristic of the fishes have been lost in modern amphibians and the epidermal scales of the reptiles have not yet developed. Rudimentary dermal scales are found buried in the skin of some caecilians, and a few toads have bony plates imbedded in the skin on the back. The spadefoot toads *(Scaphiopus* and *Pelobates)* have highly cornified areas on the feet which might be considered epidermal scales. The South African Clawed Frog, *Xenopus laevis,* has dark, cornified, epidermal structures—the so-called claws—on the first three digits of the hind foot, and some salamanders, such as *Onychodactylus,* have cornified, epidermal, clawlike structures at the ends of the digits. These are not true claws, for they lack the underpart, the so-called sole horn or subunguis, of the true claw; they are modified epidermal scales.

Digestive System

Tongue.   A well-developed definitive tongue is apparently an adaptation to life on land. Among vertebrates it first appeared in the amphibians and is characteristic of all the higher land vertebrates. A fish's food is already wet when captured and may be swallowed whole; it does not need to be moved around in the mouth for moistening or chewing before it is swallowed. The tongue of a fish is a primary tongue; a fleshy fold on the floor of the mouth that lacks intrinsic muscles and can be moved only within narrow limits. It is useful, perhaps, in pushing food farther back in the mouth for swallowing but it cannot be used to capture prey. Some aquatic salamanders (e.g., the water dog, *Necturus*) have only a primary tongue, and the tongue of some aquatic toads has degenerated and is almost or entirely absent. Other frogs and salamanders have a definitive tongue, which has, in addition to the part representing the primary tongue, an expanded glandular portion supplied with intrinsic musculature. The tongue of most frogs is attached by its base near the anterior margin of the jaw, and, at rest, is folded back on the floor of the mouth with the tip pointing toward the throat. Anyone who has seen a toad snap at a fly knows how quickly the tongue can be flipped out and how accurately it can be aimed. The tongue of some salamanders has a more extensive attachment, is mushroom-shaped (boletoid tongue), and can be shot forward to trap any unwary insect. In both groups mucous glands in the mouth provide a sticky secretion that clings to the tongue and aids in the capture of prey. Oral glands, like the definitive tongue, are a characteristic of the amphibians that is not present in lower forms.

Teeth.   With the exception of the epidermal, toothlike structures that are found in certain larval forms, the teeth of amphibians, as of most other vertebrates, are true teeth; that is, they have a hard layer of enamel surround-

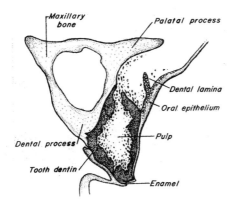

FIGURE 2-2
Section through a maxillary tooth of *Hyla cinerea*, showing its relation to the maxillary bone, oral epithelium, and dental lamina from which the new tooth is formed. [After Goin and Hester, *J. Morphol.*, 1961 (1962).]

ing the softer dentine and central pulp cavity. Toads of the genera *Bufo* and *Pipa* are toothless, but most other anurans and the salamanders have simple, cone-shaped teeth. Except for the genus *Amphignathodon* (*amphi* = both, *gnath* = jaw, *odon* = tooth), frogs have no teeth on the lower jaw. For the most part, amphibian teeth are located on the jaws, but they may also occur on bones of the roof of the mouth—the palatines and vomers. In a few genera of frogs they are also attached to the parasphenoid bones. Amphibian teeth are polyphyodont (may be replaced an indefinite number of times), and basically homodont (all the teeth along the jaw are similar). Many frogs, salamanders, and caecilians have the crown of the tooth separated from the pedicel by a zone of weakness.

Gut. The amphibian esophagus is very short, being little more than a constricted area of the alimentary tract. Usually, both the esophagus and the mouth are lined with cilia which sweep small food particles into the stomach. Secreting cells in the esophageal epithelium of some frogs produce an enzyme, pepsin, which does not begin to function until it reaches the stomach. The stomach of some salamanders is simply a straight portion of the gut, whereas in the frogs it has differentiated into a cardiac end leading from the esophagus and a short, narrow, pyloric end leading into the intestine. Digestive glands are absent from the stomachs of some fishes, but are present in all amphibians. The intestine of the caecilians has not yet differentiated into large and small intestines and shows only a slight degree of coiling. Intestinal coiling is more evident in the salamanders, and there is differentiation into a large and small intestine. In the anurans, the coiling tendency is even more marked, and the large intestine is plainly set off from the small intestine. The amphibian intestine opens into a cloaca, a common chamber into which the urinary, reproductive, and digestive systems empty. A ventral outfolding of the amphibian cloaca gives rise to the urinary bladder.

Respiratory System

By means of the respiratory system oxygen passes from some outside medium, air or water, into the bloodstream of an animal, and carbon dioxide passes from the bloodstream into the outside medium. (In a sense all respiration takes place through water, since respiratory surfaces must be kept constantly moist.) This requires some structure or structures to bring a rich supply of blood into close contact with the medium. Nowhere is the transitional position of the amphibians more clearly shown than in the variety of their respiratory processes. They have three types of highly vascular structures (ones well supplied with blood vessels) that may be used for respiration: gills, lungs, and the surface of the skin and the lining of the mouth and pharynx. As a general rule, larval amphibians use gills and the adults use lungs, but there are many exceptions. Cutaneous respiration is important in aquatic larvae and in those adults that have moist skins.

Gills.   Most fishes have internal gills which develop from tissue on the inner surfaces of the gill arches, but amphibians have external gills, which originate from the integument and are borne on the outer surfaces of the gill arches. During embryonic development, several paired pouches form in the wall of the pharynx. In most amphibians, three of these pairs break through to the outside to form gill slits. Cartilaginous supporting rods, the visceral arches, are formed in the septa between the gill slits. The gills, which consist of filaments covered with ciliated epithelium, are borne by the visceral arches anterior to the gill slits. Gills of tadpoles are usually smaller and simpler than those of salamander larvae. In one family of caecilians, Typhlonectidae, the embryos in the uterus develop two large, baglike gills from the dorsal part of the pharynx. It has been suggested that these gills function in the absorption of oxygen from the intrauterine fluid. Other caecilian embryos have three pairs of more normal, branching gills.

Among the frogs, a gill cover, the operculum, forms shortly after the external gills appear. A sheet of tissue grows backward from the hyoid region to cover the gill slits, the gills, and the region from which the forelimbs will eventually develop. It then fuses with the body behind and below the gill region to enclose the gills in an atrial chamber. This chamber has an opening to the outside—the spiracle—which may be either ventral or lateral; tadpoles of a few species have paired spiracles. Shortly after the operculum forms, the original gills degenerate and new gills develop from the walls of the gill clefts. Although they are enclosed in an atrial chamber, these are external gills derived from the integument. An opercular fold also forms in the salamanders and caecilians but it is small, consisting simply of a crease anterior to the gill region, so that no atrial chamber is formed.

During metamorphosis the gills of anurans are resorbed, the gill slits close, and the lungs take over. Salamanders show more variety. They seem to be experimenting with different solutions to the problem of breathing on land. Most terrestrial salamanders lose gills and gill slits (see Table 2-1) and acquire lungs just as frogs do. Members of the family Plethodontidae never develop lungs; the vast majority are terrestrial and depend entirely on respiration through the skin and lining of the mouth and pharynx. The aquatic *Amphiuma* and *Cryptobranchus* develop lungs and lose their gills, but retain the openings of one pair of gill slits. Adult Proteidae possess both gills and lungs, and two gill slits remain open.

Adult Trachystomata have lungs, gills, and gill slits. If the gills of a Dwarf Siren *(Pseudobranchus)* are removed, it can survive by coming frequently to the surface to gulp air. If a Dwarf Siren is put in an aquarium with a screen to prevent it from coming to the surface, its gills expand and it continues to survive. The importance of the role of cutaneous respiration in sirens and the aquatic salamanders has not been determined. The Proteidae and Sirenidae

TABLE 2-1

Distribution of Gills, Gill Slits, and Lungs
among Adult Caudates and Trachystomes

Family	Number of pairs of gills	Number of pairs of slits	Lungs
Hynobiidae	0	0	yes[1]
Cryptobranchidae			
Cryptobranchus	0	1	yes
Andrias	0	0	yes
Ambystomatidae[2]	0	0	yes[3]
Salamandridae	0	0	yes[4]
Amphiumidae	0	1	yes
Plethodontidae[2]	0	0	no
Proteidae	3	2	yes
Sirenidae			
Siren	3	3	yes
Pseudobranchus	3	1	yes

[1] Except *Onychodactylus*.
[2] Certain members of these families fail to metamorphose and retain the larval gills and gill slits when sexually mature.
[3] Reduced in *Rhyacotriton*.
[4] Except *Chioglossa* and *Salamandrina*.

are sometimes grouped as perennibranchs because both have gills as adults, but they are not closely related.

Gills of caecilians are usually resorbed before the young hatch or are born.

Lungs. The lungs of amphibians develop before metamorphosis and are relatively simple in structure. The left lung of caecilians is usually very short, with alveoli (sing. alveolus), little pockets at the ends of the bronchioles, only in the right lung. Salamander lungs are paired elongated sacs, the left one usually the longer. In some salamanders, the lining is smooth, in others alveoli are present in the basal part. Lung linings are more complex in the frogs, which is what we should expect since they have been more successful, on the whole, in making the transition from gills to lungs than have the salamanders. The walls of their lungs are made up of many folds lined with alveoli. In general, the more terrestrial the frog or toad, the larger are the alveolar respiratory surfaces in the lungs.

Respiratory Passages. Nasal passages of the amphibians lead from the external nares (nostrils) to the internal nares or choanae, openings in the roof of the mouth just inside the upper jaw. Terrestrial amphibians have mechanisms for controlling the size of the external aperture, whereas many larval salamanders, the adult Sirenidae, and frog tadpoles have valves around the internal nares to control the direction of water flow. The mouth leads into the pharynx, a gateway chamber that passes into the esophagus of the digestive system and is connected to the lungs by the trachea. The trachea is short in most salamanders but in *Amphiuma* and in the trachystome, *Siren,* it may be four or five centimeters long. Frogs generally have such a short trachea that it can hardly be said to exist, except for the modified anterior portion that forms the larynx. A trachea is definitely present in the aquatic Pipidae, in which the lungs act as hydrostatic organs. The trachea divides at the lower end into two bronchial tubes, which lead to the lungs.

During aquatic respiration water passes from the mouth to the pharynx, and out the gill slits. In salamanders and trachystomes this sets up a current of water that flows over the gills. Water enters the atrial chamber of anuran tadpoles, flows over the gills, and passes to the outside through the spiracle. Amphibians breathing air keep the mouth tightly closed and draw air in or push it out through the nasal passages by lowering or raising the floor of the mouth.

Larynx.   Since amphibians are the lowest form of vertebrates with a true ear it is not strange that they are the first ones to have the anterior end of the trachea modified to form a voice box or larynx. The larynx opens into the pharynx via a slitlike glottis, bounded on each side by a bar of cartilage that is derived from the last visceral arch. These cartilages are usually divided into upper (aretynoid) and lower (cricoid) ones. Sometimes the cricoid cartilages fuse to form a ring. Frogs and toads have two muscular bands, the vocal cords, stretching across the larynx parallel to the glottis. Air passing over the vocal cords makes them vibrate to produce sounds. Tightening or relaxing the vocal cords causes variations in pitch.

Voice is well developed in the anurans. A female frog may grunt when held in the hand or give a "mercy scream" when caught by a snake. The calling of a chorus of breeding males is one of the most familiar of all animal sounds. Some frog species may lack voices, but not all of those reported to be silent are indeed so. *Eleutherodactylus cundalli* of Jamaica, which was for many years considered a voiceless species, has been found to call only from caves and crevices.

Salamanders lack vocal cords and most of them are voiceless. Plethodontids also lack trachea and larynx. A few salamanders have "voices" but the sounds they produce seldom amount to more than faint squeaks or grunts. The Pacific Giant Salamander, *Dicamptodon ensatus,* has a low-pitched, rattling note, and makes a screaming sound when in danger. The trachystome genus *Siren* was so named because it was originally reported to "climb trees and quack like young ducklings." It does neither, but the name persists.

Vocal Sacs.   The males of many frogs have vocal sacs, outpouchings from the mouth cavity that extend ventrally and laterally under the skin and muscles of the throat. Sometimes there is a single vocal sac, sometimes a pair. In some species they are very large and extend posteriorly into the large lymphatic spaces that underlie the skin. When a frog calls, its vocal pouch is filled with air, sometimes swelling outward as a glistening bubble which collapses at the end of the call. How effective these pouches are as resonators is clear to anyone who has heard the bellowing of a Bullfrog *(Rana cates-beiana)* or the ear-splitting blast of a Great Plains Toad *(Bufo cognatus).* The calls of different species of frogs are just as characteristic as the songs of birds. Furthermore, individual frogs are apparently able to vary the pitch of the call. Some species of frogs have a "rain call" that can be easily distinguished from the mating call given at breeding ponds.

Circulatory System

In conjunction with the development of lungs, the amphibians also evolved a double circulatory system comprising a pulmonary and a systemic circulation.

Heart.   Instead of the simple two-chambered heart that is characteristic of most fishes, amphibians have a three-chambered heart with a right atrium, a left atrium, and a single ventricle. Blood returning from the systemic circulation empties into the right atrium; pulmonary veins empty into the left atrium. The lungless plethodontids lack pulmonary veins and their left atrium is small. The amphibian ventricle is not divided but nevertheless the two bloodstreams, pulmonary and systemic, mix only to a slight degree. Some amphibians have a rather complicated system of valves and partitions that helps keep the two streams separate.

Arteries.   In the primitive ancestor of the vertebrates, blood apparently left the heart through a large artery extending forward, the ventral aorta, which gave off six pairs of branches, the aortic arches. These arches curved around the gut to unite on the dorsal side and form the dorsal aorta, the main systemic artery distributing blood to the body. The history of the evolution of the arterial system of the vertebrates has been largely a history of loss of the aortic arches. Most fishes have lost the first two arches. The last four are divided into afferent and efferent parts joined by the capillaries passing through the gills. In larvel amphibians and the gilled salamanders the aortic arches are not interrupted by gill capillaries. Instead the arches give off vascular loops, which pass into the gills and branch into capillaries; these are gathered together again to rejoin the aortic arches dorsally. This simplifies the change from gill to lung respiration at metamorphosis. During this stage the vascular loops degenerate, the arches expand, and the flow of blood from heart to body continues without interruption.

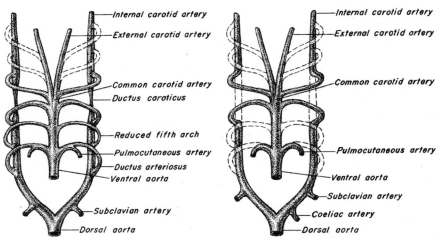

FIGURE 2-3
The condition of the aortic arches as found in most salamanders (*left*) and frogs (*right*), ventral view.

Salamanders are more primitive than frogs in the evolutionary develop-ment of their aortic arches. The first two arches are lacking; the forward extensions of the ventral aorta from which they originally arose form the external carotid arteries, which carry blood to the jaws. The third pair of arches form the internal carotid arteries which extend forward to supply the face and brain. Frequently a connection remains between the third and fourth arch. The fourth pair are the systemic arches which form the main part of the dorsal aorta. The fifth pair of arches may persist although, if so, they are much reduced in size. The sixth pair of arches gives rise to the pulmo-cutaneous arteries which carry blood mainly from the right side of the heart to the lungs and skin to be oxygenated; each arch retains its connection to one of the arms forming the dorsal aorta through a ductus arteriosus. Thus some of the unoxygenated blood passing through the sixth pair of arches may go to the body rather than to the lungs or skin. In frogs there is never any connection between the third and fourth arches, the fifth pair has dis-appeared, and the sixth arches have lost their connections with the dorsal aorta so that the unoxygenated blood they carry must all go to the lungs or skin.

Veins.   Like the other vertebrates, amphibians have an hepatic portal system. The veins carrying blood from the intestine join to form the hepatic portal vein which branches into a network of small sinuses in the liver. These vessels are then gathered together again into the hepatic veins to return blood toward the heart. The liver removes and transforms substances that have been absorbed by the blood from the intestine, storing some in the form of glycogen, returning others to the blood in a form that can be used by the cells. It also removes the harmful nitrogenous waste products of cell metabolism, changes them into harmless urea (in most adult amphibians), and empties the urea into the blood to be removed by the kidneys. Obviously, it is to the advantage of an animal to have as much blood as possible pass through the liver to be cleansed and restocked with food before its return to the heart. In this the amphibians show an advance over the fishes. Part of the blood returning from the hind legs and posterior part of the body empties directly into the hepatic portal vein through an anterior abdominal vein. The rest of the blood from the hind legs passes through the renal portal system of the kidneys. Amphibians also have pulmonary veins, which are usually lacking in fishes.

Excretory System

Getting rid of the waste products of metabolism and regulating the salt content and water balance of the body are the functions of the excretory system, which consists of the kidneys and the tubes leading from them outside the body. The excretory and reproductive systems are closely connected, especially in males, so that it is difficult to discuss one without referring to the other.

Kidney.   Three successive types of kidneys are characteristic of stages in the evolution of the vertebrates. The kidney of the adult amphibian, as in most fishes, is an opisthonephros, midway in the evolutionary series between the primitive, anterior pronephros of the hagfish and the compact, posterior metanephros of the higher tetrapods.

The kidneys of salamanders and sirens are long and are divided into a narrow anterior part and an expanded posterior part. Those of the sirens fuse posteriorly to form a knoblike "tail kidney." Caecilian kidneys are also long, extending most of the length of the body cavity, and are uniform in width throughout. Frogs have more compact kidneys that are posterior, flattened dorsoventrally, and not markedly divided into anterior and posterior parts. Caecilians and sirens have well-developed glomeruli in the anterior part of the kidney, but in salamanders and frogs the anterior glomeruli are reduced or absent. Presumably, when these glomeruli are lacking, the anterior kidney has lost its excretory function. In male amphibians, ducts from the testes enter the anterior part of the kidney and empty into the nephric ducts. These in turn join to form collecting ducts that lead into the mesonephric or Wolffian duct and then into the cloaca. Collecting ducts from the posterior part of the kidney also drain into the Wolffian duct in caecilians and sirens. The posterior kidney ducts of salamanders and frogs tend to fuse to form a ureterlike structure that may empty into the posterior part of the Wolffian duct or directly into the cloaca. This fusion is most marked in males. The Wolffian duct runs along the outside of the kidney in caecilians and salamanders but is embedded in the kidney tissue in frogs and sirens.

Bladder.   Amphibians have a thin-walled urinary bladder which opens into the cloaca and is not connected with the Wolffian ducts from the kidneys. Urine passes down these ducts directly into the cloaca and then backs into the bladder for storage. The liquid discharged by a captured toad on the hands of the collector is not urine, but reserve water stored in the cloaca. It has been suggested that the discharge lightens the animal and so facilitates its escape.

Reproductive System

As is typical in vertebrates, the amphibian reproductive system is composed of primary sex organs (gonads) and accessory organs, which include ducts and other structures that help bring the germ cells together.

Female Reproductive Organs.   The amphibian ovary is saccular with an enclosed lymphatic cavity. In vertebrates, the size of an individual ovum is largely determined by the amount of contained yolk. Amphibian eggs contain a moderate amount of yolk; more than the microscopic eggs of most mammals in which the embryo receives nourishment from the maternal bloodstream, but less than the large-yolked eggs of birds in which the young reach an advanced state before hatching. The eggs of many amphibians are about two millimeters in diameter and may be very numerous. Frogs in particular may produce thousands of mature ova at one time. Ovaries filled with ripe eggs are irregular lumpy structures that fill the greater part of the body cavity. The eggs escape into the coelom through the external wall of the ovary.

Ovaries vary in shape with variation in body form. Salamander ovaries are elongated, but are not as long and narrow as those of caecilians. Frogs, on the other hand, have decidedly shorter, more compact ovaries. The cavity of the salamander ovary is single and continuous, but in the frog ovary it is divided into a number of pockets.

Associated with the ovaries of amphibians are fat bodies or corpora adiposa. The fat bodies of salamanders are long, slender structures running parallel to the ovaries along their median edges. Those of frogs lie just anterior to the gonads and consist of several yellowish, fingerlike processes.

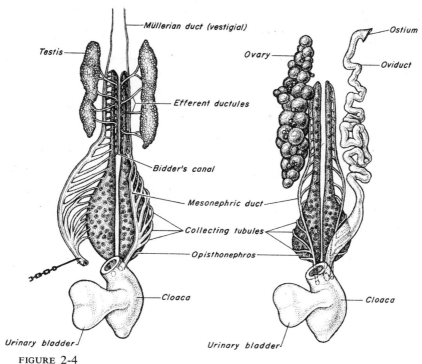

FIGURE 2-4

The urogenital organs of a salamander, ventral view: *left,* male; *right,* female.

Caecilians have leaflike fat bodies lying along the lateral edges of the ovaries. That they apparently serve as storage places for nutritive material for the developing ova is indicated by changes in size of the fat bodies during the year; they shrink as the eggs enlarge, being smallest at the close of the breeding season, after which they gradually increase in size until the maximum is reached just before the ova begin rapid growth.

The tubes that transport the products of the ovaries to the outside are the paired Müllerian ducts. These have the same general pattern for all amphibians. Anteriorly, each tube begins as an expanded, funnellike opening, the ostium, which is situated far forward in the body cavity. Eggs that have escaped from the ovaries into the body cavity are directed toward the ostium by cilia located on the peritoneum of the body wall, on the liver, and on adjacent structures. For most of its length the Müllerian tube has a thickened wall with a glandular lining. During the breeding season, the duct becomes greatly enlarged and markedly coiled, and the glandular lining epithelium secretes a clear, gelatinous substance. After entering the ostium, the egg is forced along by peristaltic waves. As it passes down the oviduct with a twisting, spiral motion, the glands deposit several layers of jellylike

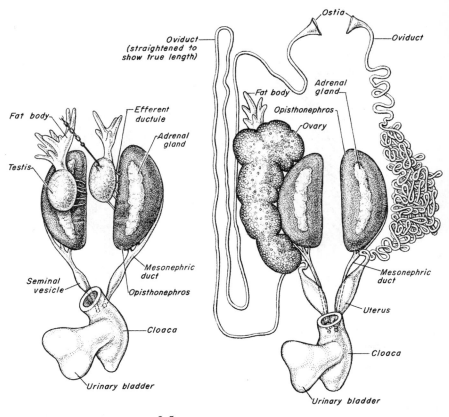

FIGURE 2-5
The urogenital organs of a frog, ventral view:
left, male; *right*, female.

material around it. Posteriorly, each oviduct is enlarged to form a uterus-like structure, which serves only as a temporary storage place for eggs that are soon to be laid. The uteri of most amphibians enter the cloaca separately, but in *Bufo*, the two uteri unite and have a common opening into the cloaca. Most female salamanders possess a dorsal diverticulum of the cloaca, the spermatheca, which serves to store sperm until the eggs are ready to be fertilized. Male amphibians usually have rudimentary Müllerian ducts.

Male Reproductive Organs.   The primary male reproductive organ is the testis, a compact structure made up of a mass of seminiferous (sperm-bearing) tubules which connect, by means of ducts, to the outside. Each testis is really a compound tubular gland. Its shape, like that of the ovary, corresponds generally to the animal's shape. The testes of primitive caecilians

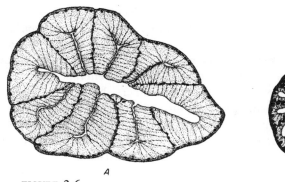

FIGURE 2-6
Cross section of oviduct of a Two-lined Salamander, *Eurycea
bislineata*, (*A*) during the breeding season (April 11), and
(*B*) after the breeding season (June 21). [From Weichert,
Anatomy of the Chordates, McGraw-Hill, 1951.]

are elongated structures that look like strings of beads. The beadlike swell-
ings, consisting of masses of seminiferous tubules, are connected by a
longitudinal collecting duct. More advanced caecilians tend to display
lobular fusion. Salamander testes are shorter and more irregular in outline.
The testes of frogs are compact and either oval or nearly rounded. There
is a pronounced difference in the size of the testes during the breeding and
the nonbreeding seasons. Like the ovaries, the male gonads have fat bodies
associated with them. They resemble the fat bodies of the ovaries in position
and appearance and, like them, fluctuate in size with the onset and passing
of the breeding season.

Male toads of the family Bufonidae have a peculiar structure—Bidder's
organ—that lies between the testis and the fat body. If the testes are removed,
these lobelike structures will develop into functional ovaries in about two
years, thus bringing about a complete reversal of sex in the male. When
this happens, the otherwise rudimentary oviducts enlarge, seemingly in
response to female sex hormones produced by the transformed Bidder's
organs. This strikingly illustrates that each sex of the vertebrates has
rudimentary organs of the other.

The efferent ductules of the salamander carry sperm from the testis to
Bidder's canal, a fine longitudinal canal that runs along the median edge
of the kidney. Bidder's canal connects with the nephric tubules in the anterior portion of the kidney by means of a number of short ducts. (Bidder's
canal is present, though in a rudimentary form, in the female as well.) The
tubules emerge from the lateral edge of the kidney to join the Wolffian duct.
In those species in which the anterior kidney has lost its excretory function
and the posterior kidney ducts enter the cloaca directly, the male Wolffian
duct is wholly reproductive in function and thus is a true ductus deferens.

The longitudinal collecting duct of the caecilian testis gives off small transverse canals that pass into the kidney and join the nephric tubules that enter the Wolffian duct. Some species show traces of a Bidder's canal.

The urogenital ducts of frogs are similar to those of salamanders, but there are minor variations. In some forms, the efferent ductules connect directly with the nephric tubules, but in others they join a Bidder's canal that is close to the median border but within the kidney. Spermatozoa are then transferred from Bidder's canal through the nephric tubules to the Wolffian duct. A Bidder's canal is present in the female, but its function, if any, is unknown. The Wolffian duct emerges from a point near the posterior end of the kidney and passes into the cloaca. Males of many species of frogs have a dilation in the Wolffian duct close to the junction with the cloaca. This forms a seminal vesicle for the temporary storage of sperm. Seminal vesicles are poorly developed in *Rana pipiens* and *Rana catesbeiana,* the two species most commonly studied in introductory zoology courses. Seminal vesicles, when present, are largest during the breeding season and shrink after breeding is over. Such seasonal changes indicate that their development and function are under hormonal control.

The Wolffian duct, particularly that of salamanders, also shows seasonal variation in size, becoming largest at the height of reproductive activity. In the nonbreeding season, the duct may be reduced to a mere thread, particularly anteriorly where it functions solely in reproduction.

Males of all known salamanders except the Hynobiidae and Cryptobranchidae undergo an enlargement of the glandular lining of the cloaca as the breeding season approaches. The glands secrete a jellylike material around a cluster of sperm to form a packet which rests on a jelly base that is also secreted by glands of the cloaca. Together they form a toadstool-shaped structure, the spermatophore.

Skeleton

Bone may develop either in the skin (the dermal skeleton or exoskeleton) or deep within the body (the endoskeleton). The endoskeleton is laid down first as cartilage which during development is replaced more or less completely by bone. Dermal bone forms directly, with no cartilaginous precursor. Generally, when we speak of the skeleton of an animal, we mean the endoskeleton, and those parts of the exoskeleton, particularly in the skull region, that have dissociated themselves from the skin and moved inward to join the endoskeleton.

The main function of the exoskeleton is to protect the body. The endoskeleton, on the other hand, provides a place of attachment for the muscles, a framework against which they can pull when moving the body. When the vertebrates moved to land, a third function of the skeleton became very important. The bodies of aquatic animals are supported by water; animals living in the much thinner medium of air lack this support. The limbs of a terrestrial animal are no longer mere steering fins—they must bear the weight of the body as it moves across the ground or through the air, and the backbone must support the internal organs. At the same time, success in the more complex terrestrial environment demands a diversity of movement unnecessary for the aquatic ancestors of the tetrapods. The skeletons of land vertebrates are more completely ossified, more solidly put together, and have more complex articulations than those of fishes. We see the beginning of these changes in the amphibians.

Except for some of the bones of the skull, the functional exoskeleton has almost disappeared in the amphibians. Certain vestiges (scales, claws) were discussed in the section on the integument.

Skull.   The skull of a modern amphibian is flattened and has fewer bones than a fish skull. The primitive cartilaginous skull box, the chondrocranium, has been partly replaced by bone. The attachment of the jaw is autostylic: that is, the upper jaw is connected directly to the skull. (The hyostylic jaw found in many fishes is braced against the skull by the hyomandibular bone which is derived from the second visceral arch.)

Vertebral Column.   The vertebral column of a fish consists of two sections: anterior to the anus are trunk vertebrae; posterior to it are caudal vertebrae. Each of the caudal vertebrae has a hemal arch, a V-shaped structure below the centrum through which pass the caudal blood vessels. The vertebral column of an amphibian shows further differentiation: the first vertebra is modified into a cervical or neck vertebra which has two concave facets for articulation with the skull, and the pelvic girdle articulates with the transverse processes of the last trunk vertebra, known as the sacral vertebra. The articulation of the pelvic girdle and the sacral vertebra still does not provide for very strong support by the hind limbs and most amphibians do not lift the body off the ground and truly walk on their legs—salamanders wriggle and frogs hop. Loss of the tail in adult frogs has, of course, involved a loss of caudal vertebrae. Their place is taken by a long, rodlike structure, the urostyle. (The structure of the individual vertebra will be discussed in Chapter 4.)

Ribs.   Ribs are present, though poorly developed, in modern amphibians. Usually they attach to the vertebrae by two heads. They are longest in caecilians, shorter in salamanders, and in most frogs are lacking entirely or are only present as cartilaginous tips on the transverse processes of the vertebrae.

Sternum. The sternum, or breastbone, which is characteristic of the higher tetrapods, appears for the first time in the amphibians, although not in a well-developed state. The ribs do not attach to it to form a definite thoracic basket. It is absent in caecilians and in some of the elongated salamanders; other salamanders have a small cartilaginous plate lying between the halves of the pectoral girdle. The sternum in primitive frogs resembles the salamander sternum but in more advanced frogs it may be rod-shaped and partly ossified.

Girdles and Limbs. The appendicular skeleton of the tetrapods comprises a pectoral girdle to which the forelimbs are attached and a pelvic girdle to which the hind limbs are attached. The pectoral girdle does not articulate with the backbone. That of the amphibians, particularly the salamanders, is largely cartilaginous rather than ossified. The amphibian pelvic girdle articulates with the sacral vertebra. It is well developed in the anurans, in which it provides attachment for some of the powerful muscles of the hind limb.

The limbs of amphibians, as of all tetrapods, are modifications of a basic pentadactyl (five-digit) plan. For the salamanders these modifications consist mainly of the fusion of some of the carpal bones and the loss of one or more of the digits. Frogs have the radius and ulna fused, and also the tibia and fibula. The bones of the tarsus are elongated to produce a hind leg that is specialized for jumping. A small additional bone, the prehallux, frequently occurs on the inner side of the foot. No amphibian has more than four fingers. Sirens are without hind limbs and pelvic girdles; caecilians lack both front and hind limbs and limb girdles.

Endocrine Glands

There is little to be said here about the structure of the endocrine glands. Most of them are found in all vertebrates and the morphological variations from one group to another seem in general to have little evolutionary significance.

Parathyroids appear for the first time as definitive structures in amphibians and seem to be concerned with controlling the level of calcium salts in the blood. They are present in all higher vertebrates and their removal results in death.

The two component parts of the adrenal gland are closely associated in amphibians as in reptiles and birds, instead of being separated as in the lower vertebrates. The reason for this difference is unknown.

Although morphologically the endocrines are of little interest, functionally they are among the most fascinating of organs. The thyroid of amphibians is the gland controlling metamorphosis. If it is removed from a tadpole, the tadpole grows into a large, fat, super tadpole, but never into a frog, although it does develop lungs and reproductive organs. If thyroid extract is fed to such a tadpole, it metamorphoses. On the other hand, if a tiny tadpole is fed thyroid extract it stops growing and quickly metamorphoses into a midget frog. The thyroid gland also controls ecdysis. An amphibian from which this gland has been removed does not shed its skin and the dead layers pile up until the animal appears much darker than normal. When fed thyroid extract it soon sheds the dead epidermal layers.

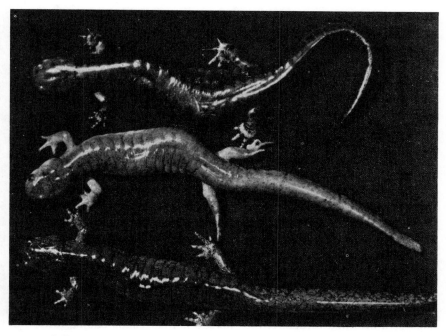

FIGURE 2-7
Small-mouthed Salamanders, *Ambystoma texanum: lower,* normal control animal; *center,* a few hours after removal of pituitary gland, pigment cells contracted; *upper,* several weeks after pituitary removal—the dark color is due to failure of the animal to shed the corneal layer of epidermis, which has become very thick. The last effect is undoubtedly caused by the failure of the thyroid gland to function in the absence of the pituitary gland. [From Weichert, *Anatomy of the Chordates,* McGraw-Hill, 1951.]

In amphibians, as in other tetrapods, the pituitary produces a number of trophic hormones that control the activity of other endocrine glands. These include both thyrotrophin (TSH), which controls the thyroid, and gonado-trophins (FSH and LH), which stimulate the production of the gametes and sex hormones by the gonads. Additional hormones of the pituitary include prolactin (LTH), which has been shown to stimulate newts to migrate to water. Increase in the production of prolactin in the spring may thus be the factor that triggers the return of amphibians to the breeding sites. Melanocyte-stimulating hormone (MSH) causes dispersal of pigment granules in pigment cells in the skin, resulting in a general darkening in those amphibians that are able to change color. Vasotocin, which is antidiuretic and increases the permeability of frog skin to water, helps regulate water balance.

Nervous System

Animals on land face a far more complicated environment than do those in water. The chief evolutionary change in the vertebrate brain is the enlargement of the cerebral hemispheres, the lateral lobes of the forepart of the brain. Primitively, the roof and sides of these hemispheres are composed of a nonnervous epithelial layer, the pallium. The amphibians have made few important advances over the fish in the structure of the nervous system, but in them for the first time scattered nerve cells appear in the walls of the pallium, foreshadowing the enormous development of the cerebrum in mammals.

Like fishes, amphibians have ten cranial nerves; higher vertebrates have twelve. Primitive fossil amphibians apparently had twelve and the reduction to ten in the modern forms seems to have resulted from a shortening of the cranium so that the eleventh and twelfth now appear as spinal nerves. The movements of tetrapod limbs are more complicated than are those of fish fins and require a more complex innervation. Enlargements of the spinal cord in the cervical and lumbar regions result; these appear first in the amphibians. The spinal cord of the salmanders extends to the tip of the tail but that of frogs is much shortened and the spinal nerves continue through the neural canal as a brushlike structure, the "horse's tail" or cauda equina.

Sense Organs

Eye.   The cornea, the clear window in the outer layer of the vertebrate eyeball through which light rays enter, becomes opaque when dried. Animals that live on land must have some mechanism to keep the cornea moist and to wash off specks of dirt. Terrestrial amphibians have developed a series of glands in the tissue around the eye, and movable eyelids to wash the secretions of these glands across the eyeball. The eyeball can be protruded or withdrawn into the eye socket. When it is drawn in, the lower lid moves up to close over it. Permanently aquatic salamanders and sirens lack eyelids, as do all amphibian larvae. Eyes are very poorly developed in caecilians and in many cave-dwelling salamanders.

Lateral Line. Fishes perceive low-frequency vibrations by means of a series of sense organs that are arranged in rows on the head and sides of the body—these are the lateral line sense organs. There is evidence that their major function is the detection of nearby objects, stationary or moving, by wave reflection. Lateral line systems are still present in larval amphibians, aquatic salamanders, and the trachystomes, but these sense organs function only in water and are absent in terrestrial amphibians.

Ear. The amphibian ear shows several advances over the ear of fishes. In addition to the sensory patches present in the membranous labyrinth of fishes, amphibians have two additional ones called the papilla basilaris and the papilla amphibiorum. The papilla amphibiorum apparently has little significance in the evolution of the ear in the higher tetrapods, but the papilla basilaris is believed to be the forerunner of the sensory structures in the cochlear duct of the reptiles, birds, and mammals.

Frogs develop a middle ear cavity from the first pharyngeal pouch. A tympanic membrane, lying flush with the surface of the head, picks up vibrations from the air and transmits them to the stapes (columella), a small rod-shaped structure lying in the tympanic chamber (cavum tympanum) behind the membrane. The stapes passes the vibrations on to the inner ear where the nerve receptors are located. A Eustachian tube connects the tympanic chamber with the pharynx and serves to equalize air pressure around the tympanic membrane. Some burrowing toads and all salamanders lack the typmanic membrane and middle ear cavity, though a stapes may be present. Vibrations seem to reach the inner ear of aquatic salamanders by way of the jaw and the bones that connect it to the cranium; terrestrial salamanders transmit vibrations via the forelimb and shoulder girdle. The "hearing" done by such means must be very limited. Perhaps this is why the salamanders have lagged so far behind the frogs in developing voices.

Jacobson's Organ. This sense organ appears for the first time in the amphibians. It is also well developed in lizards and snakes but is vestigial in most other tetrapods. In amphibians it consists of a pair of blind sacs that are connected by ducts to the nasal cavities. Since it is innervated in part by a branch of the olfactory nerve it probably plays a part in the recognition of food.

READINGS AND REFERENCES

Boulenger, G. A. *The Tailless Batrachians of Europe*. Pt. I. London: The Ray Society, 1897.

Ecker, A., R. Wiedersheim, and E. Gaupp. *Anatomie des Frosches*. 2d ed., 3 vols. Braunschweig: 1888-1904. (Old but still the definitive work on frog anatomy.)

Francis, E. T. B. *The Anatomy of the Salamander*. London and New York: Oxford University Press, 1934. (An excellent, detailed study of the anatomy of a single species of salamander.)

Holmes, S. J. *The Biology of the Frog*. 4th ed. New York: Macmillan, 1927. (A standard work on frog anatomy).

Noble, G. K. *The Biology of the Amphibia*. New York: McGraw-Hill, 1931. Reprinted by Dover, 1954. (Somewhat out-of-date but still by far the best reference work on the biology of the amphibians).

Weichert, C. K. *Anatomy of the Chordates*. 3rd ed. New York: McGraw-Hill, 1965. (All standard comparative anatomy texts include much information on the structure of amphibians.)

Young, J. Z. *The Life of Vertebrates*. 2nd ed. London: Oxford University Press, 1962. (A comprehensive, modern treatment of many aspects of the biology of the vertebrates.)

3

STRUCTURE
OF REPTILIA

Two great evolutionary advances made the reptiles more successful invaders of the land than their more primitive amphibian relatives. The first advance is the amniote egg. As the embryo develops, folds of tissue grow around it and fuse above it to form two closed sacs. The outer sac—the chorion—surrounds the whole egg with a protective membrane; the inner sac—the amnion—is filled with fluid to provide the embryo with an aquatic environment in which it can develop without danger of desiccation. A third extraembryonic membrane—the allantois—is a saclike structure that grows from the hindgut of the developing embryo and presses against the chorion; it acts as a respiratory organ and also as a storage place for metabolic waste products. A fourth membrane—the yolksac—develops around the yolk. The entire complex of developing embryo and extraembryonic membranes is surrounded by a watery albumen (except in snake and lizard eggs) and is covered with a tough shell. Since the reptilian eggshell is permeable, some environmental moisture is necessary to prevent desiccation. But the egg is much better adapted to deposition on land than the unprotected egg of even the most terrestrial amphibian.

The second advance that makes reptiles more successful than amphibians on land is the presence of epidermal scales. The amphibian epidermis is merely a soft secreting structure which offers little protection against desiccation or other hazards of life in the open air. The epidermis of reptiles has practically ceased to function as a secreting membrane and has become a protective structure, guarding the animal from actual physical injury as well as shielding it from desiccation. Epidermal scales are also found on the legs of birds and on some mammals (e.g., armadillos).

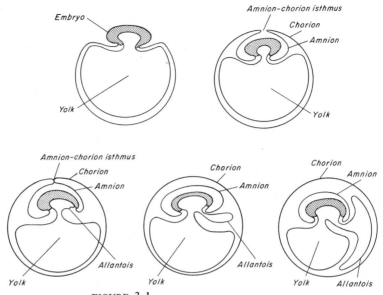

FIGURE 3-1
The formation of the amnion and chorion
in the developing amniote egg.

DEFINITION OF REPTILIA

The reptiles—turtles, lizards, snakes, ringed lizards, crocodilians, and tuatara—are amniotes that have epidermal scales but lack feathers, hair, or mammary glands. Development is direct since the need for a larval stage ended with the evolution of a shelled egg that could be laid on land. The stratum corneum is much better developed than that of amphibians and the skin is dry. In addition to heavy epidermal scales, many reptiles have bony dermal plates lying under the epidermis. Reptiles have fewer skin glands than amphibians. Respiration, as would be expected, is entirely by means of lungs, except for some that takes place through vascular tissue in the pharynx of aquatic turtles. A secondary palate, which in higher amniotes separates the nasal cavity from the oral cavity, is present though incomplete in most reptiles so that at least a partial separation occurs. In crocodilians this palate becomes complete. The regional differentiation of the vertebral column characteristic of mammals makes its appearance in the reptiles, though only the crocodilians have five clear-cut regions. There is a single occipital condyle. The legs usually have five digits, which end in true claws. The kidney is metanephric like that of birds and mammals. The heart is either three-chambered or, in the crocodilians, four-chambered. There are twelve pairs of cranial nerves. Like the amphibians, reptiles are ectothermic.

ANATOMY OF REPTILES

Present-day reptiles do not match the birds or fishes in number of species or individuals. They have, however, successfully invaded many habitats so that they now occupy the seas, fresh water, land, trees, and soil; the flying dragons *(Draco)* have even become gliders. For so small a group the reptiles show a wonderful variety of structural modifications.

Because reptiles are so diversified it is difficult to give a concise structural account that is applicable to all. The following are generalizations that may serve as a basis for understanding the modifications that characterize the groups that are discussed in later chapters.

Integument

Since the skin of a reptile does not serve as a respiratory organ, it does not need to be kept moist. The main problem however is water loss. The dry, tough, scale-covered skin of the reptiles is not completely impermeable but even so many snakes, lizards, and even turtles have adapted themselves to desert conditions and are able to go for long periods without water. (Some reptiles apparently need no other water than that obtained from their food, though most do drink when water is available.)

Epidermis. The stratum corneum of the epidermis, which first appeared in the amphibians, is much better developed in the reptiles. As with amphibians, this layer is shed periodically.

Dermis. The dermis is well developed and in many snakes and lizards has an abundance of chromatophores. That the dermis of some reptiles makes an admirable leather is attested by the popularity of alligator pocketbooks and snakeskin shoes. Poaching by hide hunters has been a major factor in the recent drastic decline of the alligator population of Florida.

Glands. The many epidermal glands that keep the skins of amphibians moist have practically disappeared in the reptiles, and the amount of water lost from the body surface is therefore much less. Most of the few remaining skin glands produce strong-smelling substances for sexual attraction. Both male and female crocodilians have two pairs of musk-secreting glands; one pair lies on the inner halves of the lower jaw and the other lies just inside the cloacal opening. Crocodilians also have a row of glands down either side of the back between the first and second rows of dermal plates; the function of these glands is unknown. Many species of lizards have series of preanal and femoral pores on the undersides of the thighs. These pores are the openings of the ducts of compound tubular glands whose secretion apparently consists of entire epithelial cells. The cells and cellular debris accumulate in the ducts to form dense secretion plugs. The glands are better developed in adult males than in females or in the young, suggesting that they play a role in reproduction, but their function remains unknown. Some snakes have glands in the cloacal opening; the nauseating odor of their secretion is probably protective. Some turtles have musk glands along the lower jaw and in the line between the plastron and the carapace; these glands are most likely used for sexual attraction. But let no man who has succumbed to the allure of *Nuit d'Amour* sneer at the lowly Stinkpot Turtle.

Scales.   The characteristic scales of reptiles are not homologous with the dermal scales of fishes but are different structures derived largely from the epidermis. However, some reptiles also have true dermal scales. They are especially well developed in turtles and fuse with each other to form the heavy plastron, or belly shell, and with the ribs and vertebrae to form the carapace, or back shell. The crocodilian dermis is thick and soft except on the dorsal side of the body and occasionally under the throat where small bony dermal scales lie beneath the epidermal scales. Crocodilians and *Sphenodon* also have gastralia, dermal bones lying in the ventral body wall between the true ribs and the pelvis. Some lizards, such as the skinks, have dermal scales (osteoderms) underlying the horny epidermal scales.

The epidermal scales of reptiles are of two sorts; one is characteristic of the snakes and lizards, the other of turtles and crocodilians.

In most lizards and snakes, each scale projects backward to overlap part of the one behind. The scales are formed from thickened areas of the integument which grow upward and backward and become cornified; they are continuous with each other at their bases (see Fig. 3-2). Periodically a new set of scales forms beneath the old and when this happens the outer, older series is shed. Snakes turn the skin inside out as they shed it; lizards simply creep out of their old skin, leaving it right side out, or shed it in flakes. Ringed lizards have narrow rings of flat scales encircling the body and tail, making them look like the segmented earthworms.

The outer portions of the so-called horns of the horned lizard *(Phrynosoma)* are modified epidermal scales as are the rattles of the rattlesnake. At the time of ecdysis the scale at the extreme tip of the tail is not shed with the others but is held in place by a bump on the newly formed scale in front. Thus a new and larger section is added to the rattle each time the snake sheds its skin (see Fig. 3-3).

Crocodilians and turtles have epidermal scales, each of which develops separately so that the scales do not form a solid sheet. Crocodilian epidermal scales cover the entire body. Instead of being shed periodically, these scales wear away and are gradually replaced.

Most turtles have large epidermal scales that do not conform in pattern to the underlying bony dermal plates of the carapace and the plastron, from

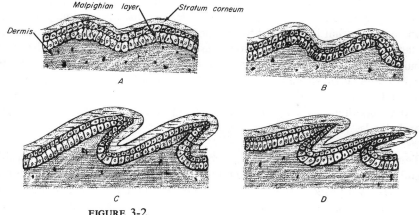

FIGURE 3-2
Diagram showing the stages in the development
of epidermal reptilian scales. [After Weichert.]

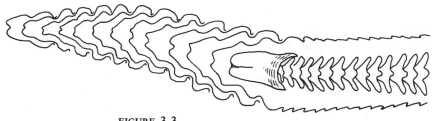

FIGURE 3-3
The structure of the rattle of a rattlesnake.

which they are separated by a thin Malpighian layer (the basal layer of the epidermis). From time to time centers in the Malpighian layer cornify to form new scales, each slightly larger than the one above it. The old scales of some turtles peel off leaving the shell smooth, whereas on others they adhere so that the new ones are topped by little mounds of older, smaller scales and the shell is rough.

Digestive System

Glands. The oral glands, which help moisten the prey to prepare it for swallowing, are better developed in reptiles than in amphibians. A palatine gland is present on the roof of the mouth. Reptiles also have lingual (tongue), sublingual (below the tongue), and labial (lip) glands. The poison glands of the poisonous snakes are modifications of the labial glands in the upper jaw.

There is one on each side which opens by a duct into the groove or cavity of the poison fang. The poisonous lizard *Heloderma* has a sublingual gland on each side that is modified to produce poison. Four ducts lead from each gland through the bone of the lower jaw to empty the poison into the vestibule in front of the grooved teeth. Thus while a rattlesnake can inject its poison at a single strike as with a hypodermic needle, *Heloderma* must hang on and chew to force the poisonous saliva into the wound. Marine turtles and crocodilians, which take their prey in water, have poorly developed oral glands.

Tongue. The tongue of turtles, crocodiles, and alligators is not protrusible and simply lies on the floor of the mouth. The Squamata have well-developed tongues (two-pronged in snakes and some lizards) that can be extended and retracted. Snakes have a small notch at the tip of the upper jaw so that even when the mouth is closed the tongue can be thrust out. The snake tongue serves solely as a sensory receptor and does not function in the manipulation of food. The chameleon tongue is the best developed among lizards; it is very useful in catching flies and other winged creatures.

Teeth. The history of the evolution of teeth has been one of reduction in number and localization of position. Fishes may have teeth almost anywhere in the mouth and in the pharynx; they may be on the tongue, the hyoid arch, and the gill arches, as well as on the jaws and roof of the mouth. Amphibian teeth are restricted to the jawbones, and to bones of the roof of the mouth—the palatines, vomers, and occasionally parasphenoids. Snakes and lizards have teeth on the palatines and pterygoids as well as on the jawbones; *Sphenodon* has vomerine teeth. Crocodilian teeth are restricted to the jawbones as are those of the mammals and the extinct toothed birds. Turtles lack teeth but instead have their jaws covered with horny beaks with sharp cutting edges (tomia).

FIGURE 3-4
The types of tooth attachment found in reptiles.

Acrodont   *Pleurodont*   *Thecodont*

Like the amphibians, the reptiles have teeth that may be replaced an indefinite number of times (polyphyodont) and that are usually the same all along the jaw (homodont). The hollow or grooved fangs of the poisonous snakes are modified teeth, the anterior teeth of some lizards are enlarged and canine-like, the lateral teeth of other lizards may be tricuspid or even oval crushing plates; these forms are thus heterodont rather than homodont. The teeth of snakes and lizards are either acrodont (attached to the upper margin of the jaw) or pleurodont (attached to the side of the jawbone). Crocodilian teeth are set in sockets in the jawbone (thecodont); this is the first evolutionary appearance of the typical mammalian tooth setting.

Gut.   The esophagus of reptiles is generally longer than that of amphibians and its walls are gathered into longitudinal folds to permit expansion during the swallowing of large prey. The stomach is clearly set off from the esophagus. Snake and lizard stomachs are long and spindle-shaped. Crocodilians have part of the stomach modified to form a muscular, gizzard-like region. The small intestine of reptiles is long and coiled. The large intestine usually has a colic cecum, a blind pouch at the juncture with the small intestine. Colic ceca are absent in crocodilians but are present in most birds and almost all mammals. The large intestine empties into a cloaca, which opens to the outside through the vent.

Respiratory System

Since reptiles as a group are the first completely terrestrial vertebrates, they are the first to depend almost entirely on the lungs to aerate the blood. Pharyngeal respiration supplements lung breathing in some aquatic turtles thus enabling them to stay under water for long periods, but in general reptiles are air breathers. Gills have disappeared completely and the reptilian lungs are better developed than are the amphibian lungs.

Among the higher tetrapods, the anterior part of the respiratory tract is separated from the anterior part of the digestive tract. The separation is initiated in the reptiles in which a hard palate begins to form. The openings of the internal nares are pushed to the back part of the mouth and the nasal passages between external and internal nares are elongated.

The reptilian larynx is generally no better developed than that of the amphibians. Most reptiles are voiceless, but some lizards produce harsh sounds and alligators bellow vociferously during the mating season. A few turtles also have voices.

Circulatory System

When they became terrestrial, the reptiles were released from the necessity of circulating blood to the gills, but their consequently increased dependence on lungs called for improvements in the pulmonary circulation and the heart.

Heart.   The heart of most reptiles is three-chambered but crocodiles and alligators have achieved a completely four-chambered heart. The right atrium, which receives unoxygenated blood from the body, is completely separated from the left, which receives oxygenated blood from the lungs. Both atria of reptiles with three-chambered hearts empty into a single ventricle, but an incomplete septum growing from the apex toward the center of the ventricle essentially separates the two bloodstreams. The septum of crocodilians, however, is complete.

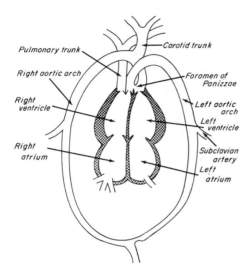

FIGURE 3-5
The relation of the aortic trunks to the heart in the alligator.

Arteries.   The third, fourth, and sixth aortic arches are present in reptiles, but their connections have been considerably modified. The truncus arteriosus, the large blood vessel that extends forward from the heart to give rise to the arches, splits at its base to form three main trunks. One of these, the pulmonary aorta, leaves the right side of the ventricle and divides into two pulmonary arteries which represent the sixth aortic arches. The left aorta (left fourth aortic arch) has become connected with the right side of the ventricle. The rest of the truncus arteriosus (right aorta) connects with the left side of the ventricle and carries oxygenated blood. In the crocodilians there is an opening—the foramen of Panizzae—between the right and left aortas as they leave the heart. The right aorta crosses over the left and runs forward on the right side. First it gives off the right (fourth) aortic arch, then passes forward and divides into the common carotid arteries. These split into internal carotids, representing the third aortic arches, and external carotids, the forward extensions of the old ventral aorta. The left and right fourth aortic arches fuse to form the dorsal aorta. Tests of the oxygen content in the blood in the various arches indicate that the left aortic arch, even though it comes from the right side of the ventricle, carries oxygenated blood. It has been suggested that since pressure is normally lower in the pulmonary circuit than in the systemic circuit, the unoxygenated blood in the right ventricle follows the path of least resistance during ventricular contraction and passes into the pulmonary aorta. The left aorta receives its blood, not from the right ventricle but via an interventricular canal, or in the crocodilians, via the foramen of Panizzae from the left ventricle. Thus all the blood entering the systemic circulation is oxygenated and the reptiles have essentially a double circulation.

Veins.   The venous system of the reptile shows little change from that of the amphibian. The large veins bringing blood back from the body are shifted over to empty into the right atrium.

Excretory System

The reptiles are the first vertebrates to have a true metanephros as a functioning kidney. The mesonephros appears only temporarily during development. During embryonic development a ureter grows forward from the posterior end of the Wolffian duct and fuses with the collecting tubules of the metanephric kidney. The Wolffian duct then carries only reproductive products and gives rise to such male structures as the epididymis, ductus deferens, and seminal vesicles.

The bladder is an outgrowth of the cloaca. Snakes and crocodilians lack a bladder but most lizards and turtles have a well-developed, bilobed structure. As in amphibians, urine passes into the cloaca and then backs up into the bladder. Some turtles have accessory bladders that are used as reservoirs for water storage.

Reproductive System

The change that, above all others, made it possible for the reptiles to become truly terrestrial was the development of the shelled, amniote egg which could be laid away from water. Surprisingly, the structural modifications in the reproductive system that accompanied this revolutionary shift in habit were relatively minor. Perhaps the most obvious evolutionary modification is the development of the copulatory organs of the males, and even these apparently are not really necessary since *Sphenodon* and most birds achieve internal fertilization without the aid of such structures.

Female Reproductive System.   Reptilian ovaries, like those of amphibians, are paired structures lying within the body cavity. The Squamata (lizards and snakes) have saccular ovaries as do the amphibians, but the other reptiles have solid ovaries as do the higher tetrapods. Reptile eggs are polylecithal (large-yolked) whereas amphibian eggs are mesolecithal (moderate-yolked). This is because the young reptile hatches at a far more advanced state of development than does the larval amphibian and hence needs more nutriment to carry it along until it is able to get food for itself. Ova size depends largely on the size of the species, but all reptilian ova are considerably larger than amphibian ova. However, since reptiles have a much smaller number (usually less than 100) of ova that mature simultaneously, the ripe ovaries, though sometimes very large, do not seem to dominate the body cavity as completely as they do in frogs. Ovaries of *Sphenodon,* turtles, and crocodilians are rather broad and symmetrically placed, but those of the Squamata are elongated. This is particularly true of the snakes, in which the right ovary lies somewhat in advance of the left, an accommodation to the snake's long, narrow body form.

Like those of the amphibians, reptile eggs break through the wall of the ovary into the celom. The oviducts (Müllerian ducts) open into the celom by narrow, slitlike ostia. Ova entering the ostium of the oviduct are forced along the tube by ciliary action and muscular contractions of the wall. Each oviduct is differentiated into regions that perform different functions in depositing the envelopes surrounding the egg when it is laid. Fertilization must take place before these envelopes are formed. Because the cilia on the walls of the oviduct beat so that the egg moves from the ostium to the uterus, there must be some mechanism to allow the sperm to travel against the main cilia-produced current in order for them to reach the ovum soon after it enters the oviduct. Turtles have a narrow band of cilia along the side of the oviduct that beat toward, instead of away from, the ostium. This is presumably the path followed by the sperm to reach the ovum. Whether or not a similar path exists in other reptiles is not known.

Sphenodon, turtles, and crocodilians possess glands in the upper part of each oviduct for secreting albumen about the ovum. These glands are lacking in the Squamata and consequently their eggs lack albumen. The lower part of the oviduct, the so-called uterus, is specialized as a shell gland to secrete the shell around the egg. The two uteri enter the cloaca independently. The oviducts vary in size with the seasons, being largest at the height of breeding activity. The right oviduct of many reptiles, particularly of snakes, is longer than the left.

Since the reptiles have a metanephric kidney with a new duct, the old Wolffian duct of the opisthonephros no longer functions in the females. It persists however as a vestigial structure, in close association with the ovary in snakes, turtles, and to a lesser extent in other reptiles.

Male Reproductive System.   Like the amphibians, the reptiles have paired testes suspended within the body cavity. Usually they lie at about the same level but in snakes and lizards one is frequently farther forward than the other. The seminiferous tubules are long and coiled. The testes may be oval, round, or pyriform; they fluctuate in size, as do those of amphibians, growing larger with the approach of the breeding season.

With the degeneration of the embryonic mesonephros the Wolffian duct no longer functions in excretion, but persists as a reproductive duct to transport sperm from the testis to the outside. The end of the Wolffian duct closest to the testis is much coiled and forms part of a tubular mass, the epididymis. Persistent mesonephric tubules are modified into efferent ductules that connect the seminiferous tubules of the testis to the Wolffian duct. These ductules form the remainder of the epididymis which may be even larger than the testis. The Wolffian duct continues posteriorly as the ductus deferens, which is sometimes straight but is more often convoluted. In most reptiles, the deferent duct on each side joins the metanephric ureter and with it enters the cloaca through a common opening at the tip of a urogenital papilla. The Müllerian duct commonly persists though generally it is reduced in size. The Müllerian ducts of the male European Green Lizard, *Lacerta viridis,* are as well developed as are those of the female. Like the gonads, the epididymes and deferent ducts of reptiles so far studied show a seasonal modification in size and are apparently under endocrine control. Many lizards undergo a periodic enlargement of some of the posterior urinary tubules of the metanephros. These enlarged tubules produce an albuminous substance which, presumably, forms part of the seminal fluid in which the spermatozoa are suspended.

Snakes and lizards have glandular structures in the cloacal walls that contribute to the seminal fluid. No accessory glands are known in the turtles.

With the development of the amniote shelled egg, internal fertilization became mandatory. This is usually accomplished by copulatory organs. Two different types of copulatory organs are found in the reptiles; each apparently represents a separate evolutionary development.

A single (occasionally multilobed), protrusible copulatory organ—the penis—is present in turtles. Only the distal end is free. There is a groove along the surface providing a passageway for the sperm. The turtle penis is apparently derived from thickened portions of the cloaca wall and is made of both connective and erectile tissue. The mass of erectile tissue is the corpus cavernosum. During mating, the corpus cavernosum becomes filled with blood thereby enlarging and stiffening the penis so that is can be extruded through the cloacal opening. This erection enables the penis to serve as an intromittent copulatory organ.

The protrusible penis of crocodilians is longer and more slender than that of turtles and the groove is deeper. A spongy structure—the glans penis—at the outer end of the groove is homologous with the mammalian glans.

Snakes and lizards have paired copulatory organs—the hemipenes (sing. hemipenis). The word means 'half a penis,' but this is not really correct since each organ is separate and complete. The hemipenes are not homologous with the true penis of turtles, crocodilians, and mammals.

The hemipenes lie on either side of the base of the tail and form distinct

FIGURE 3-6

Representative hemipenes of snakes. (*A*) Common Kingsnake, *Lampropeltis getulus*, southeastern United States; (*B*) *Coluber hippocrepis*, Italy; (*C*) Rosy Boa, *Lichanura roseofusca*, Baja California; (*D*) Puff Adder, *Bitis arietans*, Africa.

thickenings so that it is frequently possible to determine the sex of a lizard or snake without dissection. Each hemipenis is a tubular structure that can be turned inside out like the finger of a glove. It bears a groove—the sulcus spermaticus—to transport the semen from the cloaca to the tip of the hemipenis. The distal end of the organ is drawn back after copulation by means of a long retractor muscle. In a preserved specimen, the length of the hemipenis depends largely on the state of contraction of this muscle at the time of death. The external openings for the hemipenes can be seen on a snake by lifting the scale over the vent. The hemipenis of a lizard is usually short and broad and the inner surface (outer when the organ is everted) is typically pleated and folded, whereas that of a snake is longer and the surface may be covered with spines and fingerlike projections that are arranged in rosettes called calyces (sing. calyx). These apparently hold the hemipenis in the female cloaca during copulation. The hemipenis may be bilobed and the sulcus bifurcated.

In mating, usually only one hemipenis is inserted in the female cloaca. Which hemipenis is inserted is determined simply by which side the male happens to be on during copulation.

Skeleton

The skeletal modifications for terrestrial life that originated in the amphibians are further developed in the reptiles.

Skull.   The reptilian skull is not so flattened as is the amphibian skull and in general the bones are heavier and more completely ossified. Only one occipital condyle is present. Crocodilians have internal nares that open far back in the roof of the mouth and their nasal passages are separated from the mouth cavity by a bony secondary palate. (The various modifications in the temporal region will be discussed in Chapter 5.)

Vertebral Column.   A typical reptilian vertebra consists of a ventral, spool-shaped centrum and, rising above it, a neural arch that encloses the spinal cord. Small, crescentic intercentra are often wedged ventrally between the centra of successive vertebrae. On each side of the anterior face of the neural arch is a process—the prezygapophysis—that bears an articular facet; this is a smooth surface that articulates with a similar facet on a similar process—the postzygapophysis—on the next vertebra forward. Each vertebra thus articulates with the next one at three points: the centrum and the two zygapophyses on the neural arch. This is the typical tetrapod pattern. Snakes and some lizards (but not, strangely enough, the legless, snakelike ones) have two additional pairs of articular facets borne on structures at the base of the neural arch. The zygasphenes on the anterior face of a vertebra fit into grooves—the zygantra (sing. zygantrum)—on the posterior face of the next vertebra forward. This makes a strong but flexible spinal column, a decided advantage in an animal that moves by lateral undulations as a snake does.

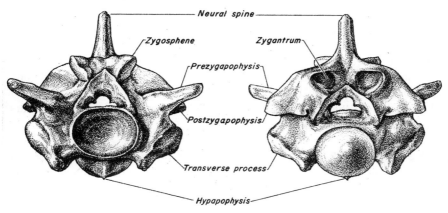

FIGURE 3-7
Vertebra of a snake, *Eunectes murinus*: *left*, anterior view; *right*, posterior view.

The first two vertebrae are modified to support the head and to allow for movement of the head on the neck. The first vertebra is called the atlas, since it bears the globe of the head much as Atlas of mythology bore the world on his shoulders. The second vertebra—the axis—permits the head to be turned from side to side.

The vertebral column of mammals is divided into five regions that are characterized by different types of vertebrae: (1) cervical vertebrae lacking free ribs, (2) thoracics with ribs, most of which attach to the sternum, (3) heavy lumbars without ribs, (4) fused sacrals attached to the pelvic girdle, and (5) caudals. Sometimes reptiles are misleadingly said to show the same divisions. All reptilian vertebrae from cervicals to caudals may have ribs. The vertebral column of snakes and legless lizards is most conveniently divided into two regions: (1) anterior to the level of the vent is a precaudal series of vertebrae bearing free ribs; (2) posterior to the vent is a caudal series with the ribs either fused to the vertebrae or absent. Usually V-shaped chevron bones, probably remnants of the hemal arches of the fishes, lie beneath the caudal vertebrae. Other lizards and *Sphenodon* have: cervical vertebrae whose ribs are usually short and do not attach to the sternum; trunk vertebrae with longer ribs, the anterior ones attached to the sternum; two sacral vertebrae with ribs attached to the pelvic girdle; and caudal vertebrae, usually with chevron bones.

Turtles have cervical vertebrae that are rather loosely articulated to allow the neck to be pulled back into the shell. The trunk vertebrae with their ribs are fused to the carapace. There are usually two sacral vertebrae that are joined to each other and connected by ribs to the pelvic girdle. Like the neck vertebrae, the tail vertebrae are not fused to the carapace.

Only the crocodilians have five easily distinguishable kinds of vertebrae in the spinal column: (1) cervical vertebrae having short ribs that are free from the sternum; (2) thoracics having longer ribs with the anterior ones attached to the sternum; (3) two to five lumbars without free ribs; (4) two sacrals attached to the pelvic girdle; and (5) a varying number of caudals.

The presence of two or more sacral vertebrae makes for better body support than is possible in the amphibians. Some reptiles are able to lift their bodies off the ground and actually walk—indeed many of the dinosaurs walked solely on their hind legs.

Ribs. Along with the variations in the vertebral columns of the reptiles goes a considerable variation in the number, structure, and attachment of the ribs. They may be attached to the vertebrae by two heads or by a single head, and they may articulate with the centra or with the neural arches. Furthermore, in a single animal the points of attachment may shift according to the position of the vertebrae in the spinal column. At their ventral ends the ribs of the anterior dorsal region usually join a sternum which is frequently cartilaginous. Trunk ribs of *Sphenodon* and lizards may join a parasternum lying between the sternum and the pelvic girdle. Turtles and snakes lack a sternum. The trunk ribs of turtles are fused to the carapace. In snakes the ventral ends of the ribs are bound by muscular connections with the abdominal scales. The posterior cervical and anterior dorsal ribs of *Sphenodon* and the crocodilians each bears a curved uncinate process which projects posteriorly to overlap the rib behind, giving strength to the thoracic body wall.

Girdles.   Usually the pectoral girdle on each side is made up of coracoid, procoracoid, and scapula, with clavicle and interclavicle frequently present. The pelvic girdle is formed of three bones on each side: a dorsal ilium that is fused to the sacral ribs and a ventral pubis and ischium. The two pubic bones join ventrally in a pubic symphysis and the two ischia also form a symphysis. Between these two symphyses is a large, heart-shaped space—the cordiform foramen. The reptile limb typically has five digits but there is a marked tendency toward a reduction of the limbs in the Squamata. Vestiges of the hind limbs remain in some primitive snakes and of the front limbs in one genus *(Bipes)* of ringed lizards. Other members of these groups lack limbs, as do a number of lacertilians.

Nervous System

The spinal cord of reptiles, like that of the salamanders, extends the entire length of the vertebral column. Cervical and lumbar enlargements, made up of the cell bodies of the neurons that go to the limbs, are present in all reptiles except in the snakes and limbless lizards. The cervical enlargements of the turtles seem unduly massive, but this is purely relative; the trunk muscles of these reptiles are so reduced that the portion of the cord between the enlargements, from which nerves pass to these muscles, is also reduced and is more slender than in other reptiles.

Some snakes and limbless lizards possess a distinct, albeit poorly developed, lumbosacral plexus, which is a network of nerves that ordinarily leads to the hind legs. This indicates that these animals must have arisen from ancestors that had legs.

Reptiles have twelve pairs of cranial nerves, as do all the higher tetrapods. The eleventh and twelfth pairs presumably represent the first two spinal nerves of the amphibians.

The cerebral hemispheres of the brain are larger than those of the amphibians, and a new area—the neopallium—appears in their roof. In the crocodilians, the migration of nerve cells to the outer wall of the neopallium results in the formation of the first true cerebral cortex.

Sense Organs

On the whole, the major changes in the sense organs that were necessitated by terrestrial existence originated in the amphibians and have simply been further developed in the reptiles.

Eye.   Snakes and some lizards lack movable eyelids but have instead a transparent window—the brille—covering the eye. It is this that gives snakes such a glassy, unwinking glare. They truly cannot close their eyes even though the eyes are always covered. Other reptiles have well-developed, movable lids, and a transparent, nictitating membrane, or third eyelid.

Ear.   The auditory part of the inner ear is more highly developed in the reptiles than in the amphibians. Most reptiles have a tympanic membrane that usually lies flush with the head, and a middle ear cavity through which sound waves are transmitted to the inner ear. Snakes lack the middle ear cavity. The bones that carry sound vibrations, instead of abutting on a tympanic membrane, connect with the jaw bones, and it is through these that the snake hears. The old saying "deaf as an adder" was probably based on the absence of external ear openings and is of course erroneous.

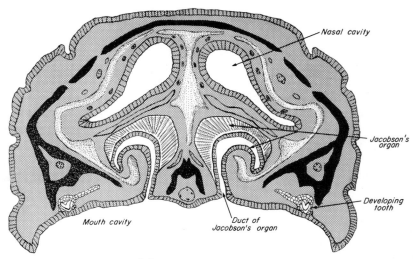

FIGURE 3-8
Transverse section through the head of a lizard,
Lacerta, showing Jacobson's organ.

Jacobson's Organ.   This vomeronasal organ is small in turtles and crocodilians. It is highly developed in snakes and lizards and no longer connects to the nasal passages but opens directly into the mouth. The flickering, two-pronged tongue of the snake picks up chemical particles from the air. When the tongue is retracted, its tips are inserted into the openings of the pockets of Jacobson's organ. The sense of smell here plays a major role in the recognition of prey, enemies, or potential mates.

READINGS AND REFERENCES

Bellairs, A. d'A. *Reptiles.* London: Hutchinson, 1957.

Carr, A. *Handbook of Turtles.* Ithaca, N.Y.: Comstock, 1952.

Gans, C., ed. *Biology of the Reptilia.* Vol. 1. New York: Academic Press, 1969.

Parker, H. W. *Snakes.* New York: W. W. Norton, 1963.

Reese, A. M. *The Alligator and Its Allies.* London: Putnam's, 1915. (The only attempt at a comprehensive survey of the features of this group.)

Romer, A. S. *Osteology of the Reptiles.* Chicago: University of Chicago Press, 1956. (An excellent, comprehensive account of the skeletal system of the reptiles, living and extinct.)

Smith, H. M. *Handbook of Lizards.* Ithaca, N.Y.: Comstock, 1946.

Smith, M. A. *The Fauna of British India. Reptilia and Amphibia.* Vols. 1–3. London: Taylor and Francis, 1931-1943. (A regional systematic study that includes a great deal of basic biology. Concise and clear accounts of structure.)

Weichert, C. K. *Anatomy of the Chordates.* 3d ed. New York: McGraw-Hill, 1965. (Or any other standard comparative anatomy text.)

Young, J. Z. *The Life of Vertebrates.* 2nd ed. London: Oxford University Press, 1962.

4

ORIGIN AND EVOLUTION
OF AMPHIBIA

During the long course of evolutionary history groups occasionally arise that make major adaptive shifts. They are able to adopt new modes of living and to move into environments that were closed to their immediate ancestors. Surely one of the most important of such shifts was that made by the amphibians, the first vertebrate animals to leave the water and spread out over the land. Indeed they were the only vertebrates to do so, for from them have come all the higher types—the reptiles, birds, mammals, and man himself. An unusual interest, then, attaches to the early evolutionary history of these relatively inconspicuous inhabitants of the land.

SOME PALEONTOLOGICAL CONSIDERATIONS

Unfortunately, amphibians as a group do not leave very good fossils. Many of them are small and have delicate skeletons, and their bones are easily scattered or crushed. This is especially true of the more recent forms, most of which are smaller than their early amphibian ancestors. And so there are great gaps in our record of amphibian evolution. Some gaps are being slowly filled as paleontologists devote more of their attention to the tiny scattered vertebrae that may be all that remains of a once numerous form. Some gaps will never be filled, and thus there will always be a certain amount of guesswork involved in our attempts to construct amphibian family trees. But the guesses of paleontologists are not blind stabs in the dark. Through studies of forms that have left clear and long-continued records, certain evolutionary principles have been derived to serve as guides. One of these is that there is a strong tendency toward irreversibility in evolution: An animal that has lost its legs does not later give rise to an animal possessing legs, and a bone that has disappeared from the skull of an ancestral form will not reappear in the skulls of descendants. We may never be able to say definitely just which fossil forms were the forerunners of the modern salamanders, but we can be quite sure that they were not animals that had completely lost their legs. Thus by a process of elimination we may be able to narrow the field to one or a few probable ancestors even for those modern animals that have no fossil history.

It is sometimes assumed that the basic criterion in the classification of animals is how recently two forms have diverged from a common ancestor. Thus, if two species are put in the same genus, their common ancestor is thought to be closer to us in time than is the common ancestor of these two and a third species that is placed in a different genus. But this criterion is really quite limited in its applicability. The length of time two forms have been separated is only roughly correlated with the degree of divergence they show. Organisms evolve at different rates. Of two lines, one may evolve very slowly and remain close to the ancestral type, the other may evolve rapidly into something quite different. Very early in the evolutionary history of the amphibians, perhaps even before they were true amphibians, the stock apparently split into several groups. One gave rise to the amphibian ancestors of the reptiles and higher tetrapods. Whether it also gave rise to any or all of the groups of modern amphibians is still a matter of lively debate among paleontologists. It is almost certain, though, that the line leading to the modern frogs had diverged from that leading to the salamanders before the birds evolved from the reptiles. But because the frogs and salamanders retain a basically similar structural pattern and mode of life, we place them in the same class, whereas the reptiles, birds, and mammals, which have departed widely from the amphibian pattern, are placed in different classes. Degree of divergence, rather than closeness of descent, is thus the basic criterion of classification.

The introduction of the factor of time brings another complication into schemes of classification. Suppose at a given period we have two genera that are obviously closely related. Each of these genera gives rise to a line, the later members of which are so distinct as to be placed in different families. Should we classify the two original genera in the families to which they gave rise, a vertical classification; or should we put them together in a third family, a horizontal classification? Either would be technically correct. Which scheme we adopt will probably depend on the completeness of the fossil record that connects the various forms, and this is more or less a matter of chance. Most classifications are actually based on combinations of the two methods.

It should be stressed that any system of classification is not an objective reality but a man-made system that is necessary because we find it difficult to study a large number of objects unless we can organize them into categories. If we knew all there is to know about all living organisms, and had a complete fossil record of every species that has ever lived, there would still be room for differences of opinion on the rank to be assigned to different groups. As it is, we know almost nothing about the detailed structure of most living species, and the fossil record is, and will necessarily remain, fragmentary. As we study more intensively the material already available, and as new forms, both living and extinct, are discovered, our ideas of the composition and relationships of the groups will shift. The classifications given in this and following chapters will almost certainly be revised in the future. They should not be taken as final or as indicating that everything is known about the relationships of the herptiles.

In studying the evolutionary history of a class it is possible to use either a horizontal or a vertical approach. With the former, we would consider all the amphibians of the Mississippian, then all those of the Pennsylvanian, and so on. This would give a clear picture of the fauna of any given period but would make it difficult to follow the various lines of descent. We shall here adopt very largely a vertical approach, following one line through its evolutionary course, then going back and picking up another line. This method involves a certain amount of mental gymnastics, of jumping backward and forward in time. To do so it is necessary to keep the geological time scale clearly in mind. Figure 4-1 shows the time scale with the known duration of the amphibian orders for reference in the following discussion.

ORIGIN OF AMPHIBIA

The earliest known animal that can definitely be assigned to the amphibians (*Ichthyostega* of Greenland) is from freshwater beds of the late Upper Devonian. Hence it is among the fishes of the Devonian or earlier times that we must look for ancestors of the tetrapods.

The primitive vertebrates of the Silurian and Early Devonian (class Placodermi) had bony dermal plates and bony internal skeletal structures. Some at least had lungs. The sharklike fishes (class Chondrichthyes), which appeared relatively late in the Devonian, had lost the ability to produce bone so that their skeletons remained cartilaginous throughout life. Presumably they had also lost their lungs, which are not present in modern members of the class. The earliest amphibians had well-developed bones and probably lungs. Bcause of both their late appearance and their loss of bone, it is highly

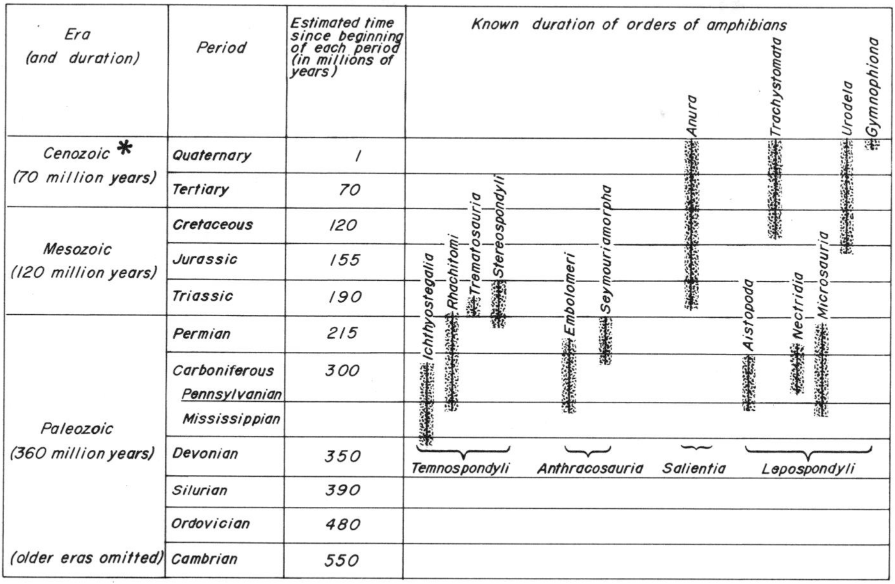

Era (and duration)	Period	Estimated time since beginning of each period (in millions of years)	Known duration of orders of amphibians
Cenozoic * (70 million years)	Quaternary	1	
	Tertiary	70	
Mesozoic (120 million years)	Cretaceous	120	
	Jurassic	155	
	Triassic	190	
	Permian	215	
Paleozoic (360 million years)	Carboniferous Pennsylvanian	300	
	Mississippian		
	Devonian	350	
	Silurian	390	
	Ordovician	480	
(older eras omitted)	Cambrian	550	

Orders (left to right): *Ichthyostegalia, Rhachitomi, Trematosauria, Stereospondyli, Embolomeri, Seymouriamorpha, Anura, Trachystomata, Urodela, Gymnophiona, Aistopoda, Nectridia, Microsauria.*

Groupings: Temnospondyli — Anthracosauria — Salientia — Lepospondyli

* The Tertiary is frequently divided into epochs. Beginning with the oldest, these are: Paleocene, Eocene, Oligocene, Miocene, and Pliocene. The Quaternary is also divided into Pleistocene and Recent.

FIGURE 4-1

The geologic time scale, showing the distribution of the orders of amphibians in time.

unlikely that the sharklike fishes could have been the precursors of the amphibians.

Ancestors of the modern bony fishes (class Osteichthyes were present in the Devonian. They were already divided into two subclasses. In one, the rayfinned fishes (Actinopterygii), the paired fins are supported only by cartilaginous rays and could hardly have developed into limbs that were able to bear the weight of an animal on land. The Actinopterygii too may be ruled out as ancestral to the tetrapods.

In many ways the other subclass of bony fishes, the fleshy-finned fishes (Sarcopterygii) are structurally close to the ancestral amphibians. In addition, present-day sarcopterygians and salamanders share a feature that sets them apart from all other vertebrates: they have very much more DNA per nucleus than have members of any other class. The African Lungfish *(Protopterus)* and several species of salamanders *(Notophthalmus viridescens, Ambystoma mexicanum, A. laterale)* have about 100 picograms of DNA per diploid nucleus. (A picogram equals 10^{-12} grams.) In contrast, forty species of actinopterygians have diploid values of from about 1 to 6 picograms per nucleus, two species of sharks have values of 5.5 and 6.5, and the single agnathan that has been reported has 5 picograms of DNA per nucleus. These figures suggest that a massive increase in amount of nuclear DNA took place in the ancestral sarcopterygians. It is possible that the presence of this excess genetic material provided the evolutionary plasticity that allowed the amphibians to make the major adaptive shift from life in water to life on land.

Apparently once the transition to terrestrial life had been made, there was a decrease in the level of DNA per nucleus. *Desmognathus monticola* and *Gyrinophilus danielsi*, the two species of salamanders with the lowest DNA/ nucleus values yet reported for the group (about 20 picograms per nucleus) are both members of the advanced family Plethodontidae. The frogs and apodans are more specialized than the salamanders. Frogs have diploid DNA nuclear values ranging from about 2 to 17 picograms and the single caecilian that has been studied has about 7.5 picograms. Diploid values reported for reptiles range from about 3 to 5 picograms, for birds from about 1.5 to 3, and for placental mammals from 5 to 7.

The sarcopterygians comprise the order Dipnoi, represented by the present day lung-fishes, and the order Crossopterygii, represented today by the coelacanth. The Dipnoi of the Middle Devonian seem to have been too specialized—especially in the structure of their teeth which were adapted to crushing shellfish—to have given rise to the amphibians.

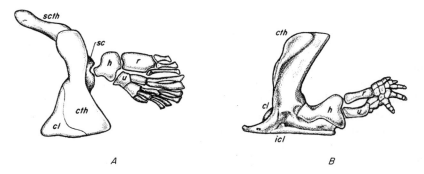

A   B

FIGURE 4-2
The shoulder girdle and pectoral fin of a crossopterygian (*A*); and the same structures of an ancient, fossil amphibian (*B*) placed in a comparable pose to show the basic similarity in limb pattern: *h*, *r*, and *u* are the humerus, radius, and ulna, respectively, of the tetrapod, and their obvious homologues in the fish fin; *cl*, clavicle; *cth*, cleithrum; *icl*, interclavicle; *sc*, scapula; *scth*, supracleithrum. [From Romer; *The Vertebrate Body*, Saunders, Philadelphia, 1955.]

The lobe-finned fishes (Crossopterygii) were abundant and diversified in the Devonian. They had lungs and their fins were supported by bony elements comparable to the bones of the tetrapod limb (see Fig. 4-2). These fins were well supplied with muscles and were probably quite mobile. Furthermore, some crossopterygians developed true internal nares. The nasal sac of other fishes is a sensory pouch quite unconnected with the mouth. In some crossopterygians, a nasal passage led from the nasal sac to an opening in the roof of the mouth. This opening—the internal naris or choana—is a characteristic tetrapod structure. While the nasal sac retains its sensory function, the passageway provides a means by which air can be taken into the mouth and lungs when only the external nostrils are above water. This probably increased the importance of the lungs as respiratory organs.

The braincase of the crossopterygians was divided transversely into anterior and posterior parts, which apparently were slightly movable on each other. No other vertebrates show this division, which may have served as a shock absorber, cushioning the brain in the hind part of the skull from jars caused by the snapping shut of the powerful jaws in the front part of the skull. Or it may have aided the fish in opening its jaws in very shallow water by allowing it to raise the upper jaw as well as drop the lower jaw. For a time the skull division was thought to debar the crossopterygians from the ancestry of the tetrapods but it has now been shown that *Ichthyostega* retained traces of a transverse division of the braincase.

The crossopterygian notochord apparently remained large. Bony neural and hemal arches were often present, but the vertebral centrum tended to be poorly ossified. Some forms had the centrum in the shape of a complete ring pierced by the notochord. In others the centrum consisted of a small bone that lay on either side just below the neural arch and a larger U-shaped bone that curved around the notochord ventrally. This is the kind of vertebra that was present in *Ichthyostega*.

Early crossopterygians had a peculiar labyrinthine infolding of the enamel of the teeth. This is not found in other fishes but it was characteristic of the earliest amphibians.

Because of the structure of their fins and vertebrae, the presence of choanae, and even the division of the braincase, the crossopterygians seem well qualified to have been the ancestors of the amphibians.

It has been suggested that these amphibian ancestors lived in warm shallow waters among thick vegetation through which they stalked their prey with well-developed, muscular fins. Oxygen content in such waters is low and their lungs probably served an important accessory breathing function. Why did they desert the water and move onto land? One suggestion is that in times of drought the animals heaved themselves onto the land to push across it in search of larger bodies of water, or perhaps to augment the limited food supply available in the shrunken ponds by feeding on the already terrestrial arthropods and on the dying fish stranded on the shores.

It has also been suggested that because these crossopterygians were fish-eaters and probably preyed on the young of their own species, small individuals most likely spent much of their time in shallow water where they could not be approached by larger fish. The young might first have moved onto land in the rush to escape a pursuing predator. Later they might have evolved the habit of staying on land at night, when humidity was higher and the danger of desiccation less, and then returning to the water to feed during the day. We shall probably never know the circumstances of that first step onto land.

CLASSIFICATION OF AMPHIBIA

Very early in their history the amphibians split into several different groups here classed as superorders. These superorders are characterized by different patterns of consolidation of the centrum (see Fig. 4–3). One line (superorder Temnospondyli) passed through a stage in which the centrum was composed of a wedge-shaped block (the hypocentrum or intercentrum), and a pair of smaller, posterodorsal blocks (the pleurocentra). The pleurocentra of the later temnospondyls were reduced and the intercentrum became the definitive centrum. The superorder Anthracosauria also passed through a stage with a compound centrum that consisted of two successive discs rather than wedges. In this line the pleurocentra fused to form the main body of the centrum and the intercentrum was lost or persisted only as a vestige. This was the line that gave rise to the amniotes. The centra in the remaining two superorders (Salientia and Lepospondyli) show no evidence of having passed through a double stage in their evolution. The centrum of the Lepospondyli is a simple, spool-shaped structure, which apparently ossifies directly around the notochord without being laid down first as cartilage. That of the Salientia forms through chondrification and then ossification of part or all of the perichordal tube (sheath around the notochord) below each neural arch.

The position of the modern orders of amphibians is a matter of lively debate among paleontologists. The recent forms are very different but they share certain features that have persuaded many workers that the modern amphibians evolved from a single Paleozoic stock and should be placed together in a subclass Lissamphibia. The presence of a zone of weakness separating the crown and pedicel of the tooth in frogs, salamanders and caecilians is one character that is used to support their inclusion in a single subclass. A similar tooth construction has recently been reported for a member of the Temnospondyli.

Another character shared by many salamanders and frogs is the presence of two middle ear bones—the stapes and the operculum—that fit into the oval window of the inner ear. The stapes is homologous with the hyomandibular, the bone in fishes that abuts on the otic region of the skull and serves to prop the jaw against the skull. The operculum has not been definitely identified

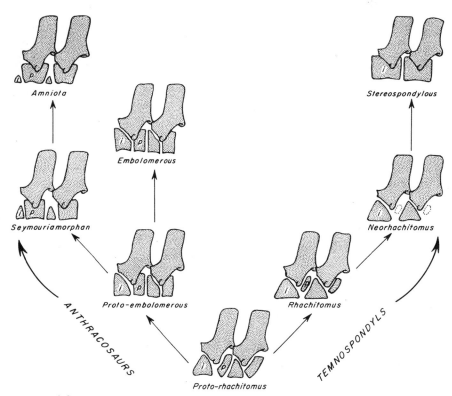

FIGURE 4-3

The phylogeny of vertebral types; *i*, intercentrum; *p*, pleurocentrum (true centrum). Viewed from left side. [After Romer, 1966.]

in any other group, fossil or recent, but—because in modern amphibians it is frequently cartilaginous, or fused to the stapes, or both, so that its presence can be determined only by studying developmental stages—its absence from the fossil record is not surprising. The stapes develops first in salamander larvae and seems to function in the transmission of waterborne sound vibrations from the jaw to the inner ear; this mechanism would be of little use on land. At metamorphosis, the operculum appears. It is connected by an opercularis muscle to the shoulder blade and apparently transmits vibrations that pass from the ground through the forelimb. At metamorphosis, most frogs develop a tympanic membrane that is fastened to the distal end of the stapes and housed in a notch—the otic notch—at the rear of the skull. A middle ear cavity forms around the stapes. Airborne vibrations are transmitted via the tympanum and stapes to the inner ear. The operculum and opercularis muscle are also present and now seem to help the frog maintain equilibrium on a tilted surface.

A notch similar to the otic notch was present in the crossopterygian ancestors of the amphibians but in them it apparently represented the opening of the first gill pouch; the hyomandibular retained its connection with the jaw. Among the fossil amphibians an otic notch was present in the ichthyostegids, temnospondyls, and anthracosaurs. It is not clear from the fossil record just when the tympanic membrane appeared or when the hyomandibular was converted into a stapes. An otic notch was absent in most fossil lepospondyls and is lacking in all modern amphibians except frogs. It is frequently assumed that the salamanders descended from forms that had a middle ear construction similar to that of the frogs and have lost not only the otic notch but also the tympanum and middle ear cavity. Another explanation seems more plausible to us. It is highly unlikely that the protoamphibian that first moved out on land had already developed a mechanism for receiving airborne vibrations, but the animal was probably capable of receiving waterborne vibrations much as salamander larvae apparently do today. It may be that the first structure that evolved for hearing on land was the operculum. Its retention in this capacity by salamanders would then represent a primitive state. The development of the tympanum and middle ear cavity provided a more efficient hearing mechanism, and the operculum was either retained as an aid to equilibrium, as in frogs, or lost, as in the higher tetrapods.

We are here tentatively placing the salamanders, trachystomes, and caecilians with the lepospondyls and separating the frogs as a superorder Salientia, which is perhaps allied to the temnospondyls.

Superorder Ichthyostegalia

This superorder includes the most primitive of all amphibians. *Ichthyostega* was a fairly large animal with a skull 150 mm or more in length. It seems to have been intermediate between a fish and a tetrapod in appearance. It had short, stubby, pentadactyl limbs instead of fins, and a tail somewhat like that of a modern lungfish with a caudal fin supported by fin rays. It had a short, rounded snout and a relatively long skull table (the posterior part of the skull roof). In these proportions it was intermediate between the crossopterygians and the later amphibians, in which the snout region is well developed and the skull table reduced. Lateral line canals were present, suggesting that these aquatic sense organs were still important to *Ichthyostega,* and there was a single occipital condyle—another fishlike character. The internal nares were bounded laterally by a slender process from the maxilla rather than by a broad bar of bone as in most tetrapods. The centrum of *Ichthyostega* was like that found in some crossopterygians, with a U-shaped ventral portion and a small block on either side just below the neural arch. *Ichthyostega* must have spent most of its time in the water but was almost surely able to crawl onto land.

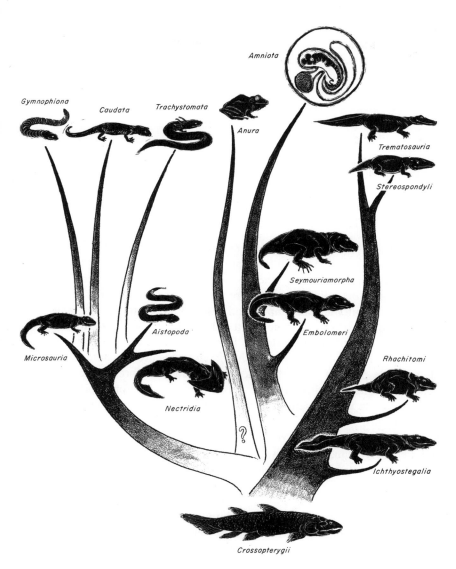

FIGURE 4-4
The phylogenetic arrangement of the orders of amphibians.

Besides *Ichthyostega,* the superorder includes several other genera of Devonian and Carboniferous amphibians. Some of them, at least, retained an armor of fishlike, bony, ventral scales, and also had traces of scales on the back and flank. Although very primitive in most respects, they all showed specializations indicating they were side branches rather than a part of the main stem of amphibian evolution. Thus, the intertemporal bone of the skull, which was present in other primitive amphibians, was lacking in all the known genera of ichthyostegalians.

Superorder Temnospondyli

The Temnospondyli were widespread and flourishing during the later Paleozoic. The earliest members of the group were aquatic but some of the later forms were well adapted to life on land. Then, in the late Permian and Triassic, the temnospondyls returned to life in the water. The line died out at the end of the Triassic.

Order Rhachitomi.   The rhachitomes of the Carboniferous and Permian were the most abundant of all fossil amphibians in number of genera. Almost all the large amphibians of the Permian belong in this order. The more primitive members of the group were probably largely aquatic but some of the later rhachitomes were apparently well adapted to life on land. Each neural arch had a centrum composed of interlocking, wedge-shaped blocks— the typical rhachitomous vertebra. It has recently been suggested that the intercentrum and pleurocentra of the rhachitomes are not homologous with the separate parts of the ichthyostegid vertebra but represent new, tetrapod structures. The occipital condyle was single, double, or triple; the intertemporal bone was present in early members of the group but disappeared in later forms.

The loxommids, a group of primitive forms from the Upper Mississippian and Pennsylvanian, are known only from skull elements. Whether or not they had rhachitomous vertebrae, their skulls were in many respects similar to what we should expect to find in the early rhachitomes. They were, however, specialized in having unusually large, elongate orbits. They were probably an early side branch of the rhachitomous stock.

Unquestioned primitive rhachitomes were present in the Pennsylvanian and had developed considerable morphological variation. Some were very long faced, others were short faced. The latter may have been specialized or may simply have been larval or even neotenic forms.

The typical rhachitomes of the Permian represent the culmination of the development of terrestrial life in the temnospondyls. Many of them were quite stout limbed and seem to have been well adapted for life on land. Some were giants among the amphibians—*Eryops* of the Lower Permian was about 150 cm long.

Order Trematosauria.   The trematosaurs were a spectacular group that was probably derived from the later rhachitomes. They are known only from the Lower Triassic. Their triangular skull was rather high and narrow, sometimes with a very elongated snout. Grooves indicate the presence of lateral line sense organs. The intertemporal bone was absent, and there were two occipital condyles. The rest of the body is poorly known, but was apparently long and slender and had feeble legs. The vertebrae were still basically rhachitomous but the pleurocentra were reduced and sometimes remained cartilaginous.

It is apparent that the trematosaurs were aquatic in habit and their sharp-pointed teeth indicate that they ate fish. In Greenland and Spitzbergen their fossils occur in beds with abundant fish remains. Since these beds are, in part, marine, it may be that the adults were adapted to enter the sea to feed but returned to fresh water to breed. There are indications that in the early Mesozoic the actinopterygian fishes started moving from fresh water to salt water and it may be that for a short time the trematosaurs went along a parallel line of evolution. But the trematosaur experiment was not a success, and they became extinct by the end of the Triassic.

Order Stereospondyli.   These last and most degenerate members of the Temnospondyli had given up their foothold on land and slipped back to an aquatic, mainly bottom-dwelling existence. The body was flattened and the limbs were too small and weak to have supported the animal on land. Lateral line canals were present on the skull. The intertemporal was absent, and the occipital condyles were double. The pleurocentra were reduced, seldom ossified, and often entirely absent. The intercentra were always highly developed and sometimes formed complete discs around the notochord. The stereospondyls evolved from the rhachitomes in the Permian and were rather common in the Triassic.

The dividing line between the rhachitomes and the stereospondyls is not a sharp one; the group known as neorhachitomes might just as well be placed at the top of the former as at the bottom of the latter. Like all the stereospondyls, the neorhachitomes were characterized by features that were associated with a degenerative trend toward an aquatic existence. Their skulls were flattened and marked with lateral line grooves; many of the elements of the braincase were either unossified or only partly developed; the limbs were small with poorly ossified bones and the shoulder girdles were greatly expanded and ventrally flattened. The neorhachitomes lived in the late Permian and early Triassic.

The long-snouted capitosaurs of the Triassic, which were obviously derived from the neorhachitomes, form a rather compact group showing little variation. Some are notable for their size. *Mastodonsaurus,* with a skull length of over 90 cm, was the largest of the amphibians.

The remaining stereospondylous amphibians of the Triassic are known as the short-faced stereospondyls. Their skull was extremely broad and flat; their face was relatively short, but the part of the skull behind the eyes was decidedly elongated. The external nares lay close together and were much enlarged (it is possible that they accommodated the tips of the tusks of the lower jaw). Like the capitosaurs, the short-faced stereospondyls apparently evolved from the neorhachitomes and became extinct at the end of the Triassic.

Superorder Lepospondyli

This group of small amphibians, which flourished in the pools of the Carboniferous, had the centrum of the vertebra formed as a bony cylinder around the notochord instead of as separate intercentral and pleurocentral blocks. They were all animals of modest size; many of them had lost their limbs and were eellike in appearance. These Paleozoic lepospondyls are divided into three orders: Aistopoda, Nectridia, and Microsauria. Three modern orders of amphibians (Trachystomata, Caudata, and Gymnophiona) have vertebrae that are apparently constructed on the same basic plan. Although there are great gaps in the fossil record between the earliest known members of these orders and the lepospondyls of the Paleozoic, they may be members of the same stock and are included here in the same superorder.

Order Aistopoda. This order includes only a small number of elongate, limbless, possibly fossorial amphibians. Some were snakelike and had more than a hundred vertebrae. They appeared in the Mississippian and disappeared in the early Permian.

Order Nectridia. There is some question about whether the aistopods and nectridians should be placed in different orders since they show many similarities. The nectridians were somewhat more varied, but all had the neural and hemal arches of the caudal vertebrae expanded into fanlike structures. Very common in the Pennsylvanian pools were a number of small, slender, ell-shaped nectridians with long, pointed heads and with limbs either reduced or absent. Also abundant in the Pennsylvanian, and lasting into the Lower Permian, was another group of nectridians. They were somewhat larger (some reached lengths of about 600 mm) and had flattened bodies, reduced limbs, and weird-looking heads in which the posterior corners of the skull were drawn back into a pair of grotesque horns. They were probably bottom dwellers, and swam by skate-like undulations of the body.

Order Trachystomata.   This order includes two modern genera—*Siren* and *Pseudobranchus*—and a number of extinct forms that go back to the Cretaceous. They are usually classified with the Caudata, but they differ sharply from all known salamanders and, in the structure of their vertebrae, show resemblances to the aistopod-nectridian stock. Like these fossil amphibians, trachystomes are eel-shaped and have expanded arches on the tail vertebrae. The hind limbs are absent, the fore ones minute. (Since the pectoral girdle is not attached to the rest of the skeleton, the tiny limb bones usually become separated from the rest of the body shortly after death. Limbs are not known for most fossil trachystomes, but their presence in the modern forms shows that front legs must also have been present in the ancestral species and raises the question of whether some of the supposedly legless aistopods may not also have had vestigial forelimbs.)

Order Microsauria.   The microsaurs were small creatures that in general looked more like modern salamanders than did any of the lepospondyls previously mentioned. They varied considerably in structure; some paralleled the aistopods and nectridians in the reduction of the limbs but most seem to have had a more normal tetrapod build. Together with the aistopods and nectridians, they flourished during the Carboniferous; indeed, the lepospondyls seem to have been the most abundant amphibians in the swamp pools of that time. Microsaurs became extinct during the Permian.

Order Gymnophiona (Apoda).   The caecilians are slender, wormlike creatures that are practically blind and lack limbs or limb girdles. Their compact skulls are modified for burrowing and lack some of the bones present in primitive amphibians. No fossil caecilians have ever been found to bridge the gap between the modern ones and their possible Paleozoic ancestors.

Order Caudata.   Modern salamanders are mostly small animals with long slender bodies and tails, and rather feeble legs. Like other modern amphibians, they show a strong tendency toward a reduction in the bones of the skull. They have been traced as far back as the Jurassic, but there is still an enormous gap in the record between them and the lepospondyls of the Paleozoic. The aistopods and nectridians, though, were even then very specialized; it is possible that we should look for the ancestors of the Caudata among the microsaurs.

The scattered salamander vertebrae from the Jurassic can not be surely assigned to any of the modern families. Two Cretaceous genera are placed in the extinct family Prosirenidae. Vertebrae from the Upper Cretaceous and Paleocene have been described as belonging to an extinct family, Scapherpetontidae. Another family, Batrachosauroididae, has been erected for two genera, *Batrachosauroides* and *Opisthotriton,* that are known from the Eocene and Miocene. Other fossil salamanders have all been assigned to modern families (for descriptions of these families, see Chap. 13). Of these, the Amphiumidae and perhaps Proteidae go back to the Cretaceus, the Ambystomatidae and Salamandridae to the Paleocene, and the Cryptobranchidae to the Oligocene. One large Miocene cryptobranchid, *Andrias scheuchzeri,* was described in 1726 as the mortal remains of a human sinner drowned by the Noachian Deluge, and named "Homo diluvii testis" (man witness of the flood). This, of course, was in the days before Darwin, when it was popular to explain away all fossils as having been left over from the Flood.

The Plethodontidae have not been positively identified before the Miocene. The Hynobiidae are as yet unrecorded as fossils.

Superorder Salientia

The frogs, with their elongated hind legs and shortened backbones, are among the most specialized of all vertebrates. At present they cannot be traced to any group of Paleozoic amphibians. Zoogeographic evidence suggests, however, that this may be true in part simply because the appropriate fossil beds, viz., the extensive carboniferous beds in Antarctica, have not yet been fully explored.

The earliest fossil that might be assigned to some salientian ancestor are tracks in the Ecca Formation of the basal Permian in South Africa. These prints are of the fore feet and suggest that the creature swam about and groveled on the bottom.

Order Proanura.   *Triadobatrachus (Protobatrachus)* from the Lower Triassic of Madagascar may represent either an intermediate stage in the evolution of frogs, or a striking case of parallel evolution, or a metamorphosing individual in which the tail had not been resorbed and the sacroiliac articulation was not yet formed. Its froglike skull and long hind legs align it with the Salientia. Until more and better material is found and studied it is retained in the order Proanura as a discrete family, Triadobatrachidae.

Order Anura.   The remaining Mesozoic salientians are all referable to the order Anura. There are three basic groups of frogs recognizable in the Jurassic and Cretaceous beds and they all have in common that they are members of primitive, aquatic groups. These groups are the ascaphoids, discoglossoids, and the pipoids.

The ascaphoids are represented by *Vieraella* from the Lower Jurassic and *Notobatrachus* from the Middle Jurassic of Patagonia. They may be related to the modern ascaphids but, pending the discovery and study of additional material, they are retained in a family of their own, the *Notobatrachidae.*

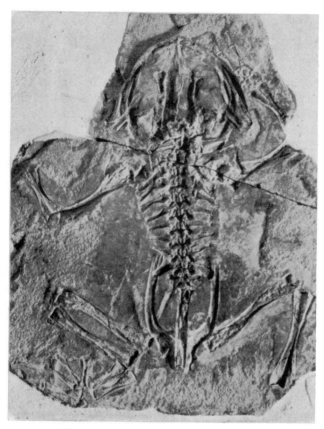

FIGURE 4-5
Notobatrachus degiustoi Reig, a primitive frog from the
Jurassic of Patagonia. [Courtesy of Professor Osvaldo Reig.]

The discoglossoids first appear in the Upper Jurassic of Europe in the form of *Montsechobatrachus* and *Eodiscoglossus*. The European forms were formerly set off as the family Montsechobatrachidae but they probably represent nothing more than early representatives of the family Discoglossidae.

The pipoids can be traced back with some degree of surety to *Thorachialicus* and *Cordicephalus* of the Lower Cretaceous of Israel. It seems likely that *Eobatrachus* from the Upper Jurassic of Wyoming, *Eoxenopoides* from the uppermost Cretaceous of South Africa, *Saltenia* from the uppermost Cretaceous of Argentina and *Shelania* from the Paleocene of Patagonia belong in the pipids also. All of these Mesozoic and Paleocene forms are included in the Recent family Pipidae.

Either in the late Mesozoic or very early Cenozoic a sideline of pipoid frogs developed into what is now recognized as the family Paleobatrachidae. Comprising only two genera and five species, it was an ill-fated group and became extinct by the Middle Tertiary.

The remaining fossil frogs have all been placed in families that have living representatives. As pointed out above, the discoglossids and pipids go back to the Jurassic. The Ranidae, Leptodactylidae, Bufonidae, Pelobatidae, Rhynophrynidae, and Pelodytidae had all appeared by the Eocene, and the Hylidae and Microhylidae by the Miocene. The remaining families have not been reported from the Tertiary, but this is undoubtedly because of the incompleteness of our knowledge. It may someday be shown that most modern frog families go back to Mesozoic times.

Superorder Anthracosauria

This superorder includes the members of that line of amphibian descent in which the pleurocentrum came to form the main body of the centrum while the intercentrum was progressively reduced. Anthracosaurs appeared first in the Mississippian and then, as amphibians, became extinct at the end of the Paleozoic. But before they disappeared, they gave rise to true terrestrial vertebrates, the reptiles.

Order Schizomeri.   The most primitive known anthracosaur, the Mississippian *Pholidogaster,* had a large, crescent-shaped intercentrum and pleurocentra that formed paired half rings. The skull resembled that of later anthracosaurs but the elongated trunk and reduced limbs indicate that this animal had already diverged from the line leading to the reptiles.

Order Embolomeri.   These were common amphibians of the Carboniferous, primitive in many ways, yet in some respects already foreshadowing the reptiles. Both the intercentrum and the pleurocentrum were complete discs. There was a single occipital condyle and the intertemporal bone was still present. The later forms, at least, had long bodies, powerful tails, and small limbs, and some had very elongated snouts. They were almost certainly aquatic, and were probably fish eaters. Some were very large; *Pteroplax* reached an estimated length of more than 450 cm. The embolomeres became extinct in the Lower Permian.

Order Seymouriamorpha.   The seymouriamorphs flourished from the Upper Pennsylvanian to the Upper Permian. They were mostly moderate in size, stockily built, had stout limbs, and were certainly capable of walking on land. *Diadectes* reached a length of 3 meters. The group takes its name from what is probably the most widely known fossil amphibian, *Seymouria.*

A number of reptilelike characters were present in the group. Among these were five digits in the front foot, swollen neural arches, and a true pleurocentrum in combination with a reduced, crescent-shaped intercentrum. Seymouriamorphs also had such nonreptilian characters as an intertemporal bone in the skull, traces of lateral line grooves, more bones in the lower jaw than are found in reptiles, a single sacral vertebra, and an enormous otic notch. The stapes in amphibians is directed dorsally and the otic notch lies high on the skull. The stapes of the reptiles has shifted in position and now points down and back toward the angle of the jaw. The amphibian otic notch is lost in the primitive reptiles, though a somewhat similar notch evolves later in some reptilian lines.

Seymouria is so nearly intermediate in structure between the amphibians and reptiles, that it has sometimes been classified as the most advanced of the former, sometimes as the most primitive of the latter. Some of the fossils assigned to the group, however, seem to represent aquatic larval stages. This indicates that the seymouriamorphs did not lay shelled, amniote eggs on land, and that they probably should be classified with the amphibians. Thus we must look to another group for the ancestral reptiles.

Order Diplomeri.   This is a group of rather poorly known anthracosaurs from the middle Pennsylvanian that seems almost ideally intermediate between the amphibians and reptiles. *Gephyrostegus* and its allies retained the amphibian otic notch and intertemporal bones, but in the rather high, narrow skull, the structure of the atlas and axis, and the general body configuration they are closely similar to the most primitive reptiles. They are too late in time to be the direct ancestors of the reptiles, but they almost surely represent the line from which the reptiles sprung.

READINGS AND REFERENCES

Goin, O. B., and C. J. Goin. "DNA and the Evolution of the Vertebrates." *The American Midland Naturalist,* vol. 80, no. 2, 1968.

Griffiths, I. "The Phylogeny of the Salientia." *Biological Reviews of the Cambridge Philosophical Society,* vol. 38, no. 2, 1968.

Nevo, Eviator. "Pipid Frogs from the Early Cretaceous of Israel and Pipid Evolution." *Bulletin of the Museum of Comparative Zoology at Harvard College,* vol. 136, no. 8, 1968.

Parsons, T. S., and E. E. Williams. "The Relationships of the Modern Amphibia: A Re-examination." *Quarterly Review of Biology,* vol. 38, no. 1, 1963.

Piveteau, Jean, ed. *Amphibiens, Reptiles, Oiseaux.* Traité de Paléontologie, vol. 5. Paris: Masson, 1955. (A well illustrated, modern treatment of the fossil amphibians and reptiles.

Romer, A. S. *Vertebrate Paleontology.* 3rd ed. Chicago: The University of Chicago Press, 1966. (Includes a good summary of amphibian evolution, less technical than the other references given here.)

Szarski, Henryk. "The Origin of the Amphibia." *Quarterly Review of Biology,* vol. 37, no. 3, 1962.

Watson, D. M. S. "The evolution and origin of the Amphibia." *Philosophical Transactions of the Royal Society of London,* ser. B, vol. 215, 1926. (In this and other papers Watson laid the foundation for the modern classification of amphibians.)

5

ORIGIN AND EVOLUTION
OF REPTILIA

Amphibians were like the early explorers of the New World who still called the Old World home. They invaded and explored the land but most could not travel far from their aquatic homeland and necessarily returned to it to reproduce. The reptiles were the true colonizers who settled down to live and reproduce in the New World of dry land.

THE CLEIDOIC EGG

The development of the cleidoic (or closed) egg marked the dividing line between the amphibians and the reptiles. Many present-day amphibians (caecilians, a few salamanders, a number of frogs) lay their eggs away from water (see Chap. 6). Some produce eggs that are relatively large yolked and the young, instead of going through a larval stage, hatch in the adult body form. Such eggs need only shells and extraembryonic membranes to make them typical reptilian eggs. Most of the amphibians that lay terrestrial eggs live in mountainous regions where there are few large open bodies of undisturbed water to which they may resort for breeding. It is possible that the reptilian egg evolved as an adaptation to mountainous conditions. At any rate, the first animals that had reached the morphological level of true reptiles appeared in the Pennsylvanian, a time of mountain building.

It has recently been suggested that the first of the amniote extraembryonic membranes to evolve was the allantois, which develops as an outgrowth of the posterior part of the embryonic gut. Reptiles presumably evolved from amphibians that already laid a small number of large-yolked eggs beside the water. The large amount of yolk allowed a prolonged developmental period and eliminated the need for an aquatic, feeding larva. A longer developmental period, though, exposed the embryo to an increased danger of desiccation. Water conservation became a problem. Much of the water loss during development results from the need to dispose of the nitrogenous waste products of metabolic activity. Aquatic larval amphibians excrete ammonia, which is both highly toxic and highly soluble. Embryos developing in eggs laid on land could not afford to expend the large amounts of water needed to remove the ammonia from the body. It is suggested that the embryos of the amphibian ancestors of the reptiles excreted urea, which is much less toxic than ammonia, just as terrestrial adult amphibians do today. Solutions of urea have a high osmotic pressure and, if retained within the egg, might have helped the embryo to absorb water from the environment. This in turn could have led to the development of an enlarged storage sac (allantois) at the hind end of the gut to hold both the accumulated water and dissolved urea. The water could then be resorbed by the allantoic blood vessels. As the allantois increased in size it would have spread between the layers of extraembryonic tissue, forcing them into folds around the embryo. When the folds met and fused, they formed the two other extraembryonic membranes, the amnion, which surrounds the embryo proper, and the chorion, which surrounds the embryo, amnion, allantois, and yolk sac. This is not the way the membranes develop in the embryos of modern reptiles, but it may well be the way they first evolved (see Fig. 5–1).

THE REPTILE SKULL

The first true reptiles were the Captorhinomorpha. The captorhinomorph skull was solidly roofed over by dermal bones. These bones were arranged in three rows on each side: a median row that included the frontal and parietal bones; a lateral row that comprised the postorbital and squamosal bones; and a marginal row in which the jugal and quadrate bones were major elements. The otic notch had disappeared, except perhaps in *Romeriscus,* the earliest known genus. The brain was very small and was encased in a narrow, bony box—the braincase—that lay beneath the center of the skull roof. The muscles of the jaw attached to the underside of the roofing bones. During the evolution of the reptilian skull, openings developed between the bones of the once-solid skull roof. If the openings developed along the sutures between bones of the median and lateral rows, a pair of dorsal tem-

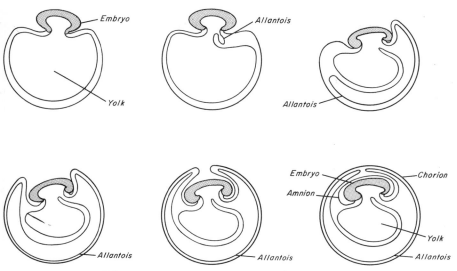

FIGURE 5-1
Hypothetical origin of the extraembryonic membranes of the reptiles.

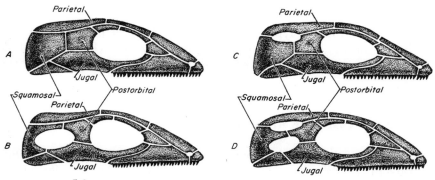

FIGURE 5-2
Diagrammatic view of reptilian skulls to show variation in temporal openings: (*A*) anapsid, the primitive type without openings; (*B*) synapsid, with a lateral opening, the type found in the ancestors of mammals; (*C*) euryapsid, with an upper temporal opening, the type found in plesiosaurs; (*D*) diapsid, the type found in *Sphenodon*, the crocodilians and other archosaurs, and in modified form in the lizards and snakes.

poral openings formed; if they developed between the bones of the lateral and marginal rows, a pair of lateral temporal openings appeared. In some reptiles both dorsal and lateral openings developed (see Fig. 5–2).

The subclasses into which the reptiles are divided are defined, in part, by the nature of the temporal openings in the skull. Early workers paid most attention to the arches, the bars of bone that lie below the openings. This is

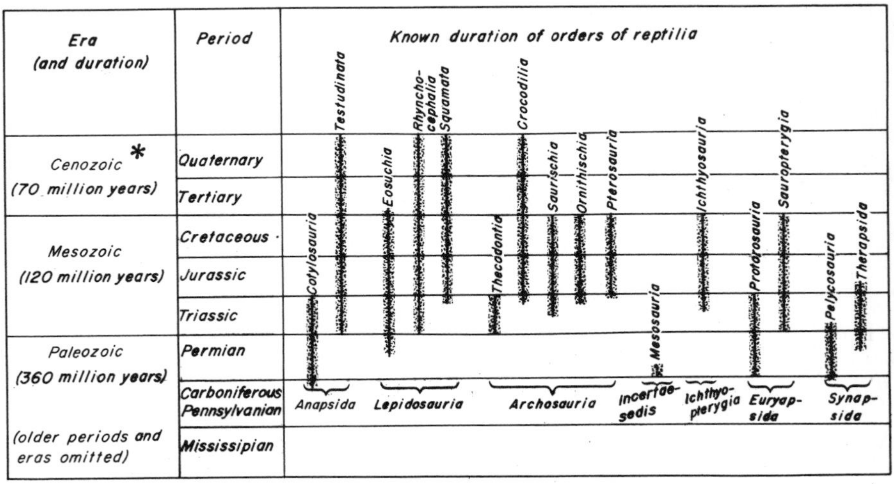

Era (and duration)	Period	Known duration of orders of reptilia
Cenozoic * (70 million years)	Quaternary	
	Tertiary	
Mesozoic (120 million years)	Cretaceous	
	Jurassic	
	Triassic	
Paleozoic (360 million years)	Permian	
	Carboniferous Pennsylvanian	
(older periods and eras omitted)	Mississippian	

Orders (diagonal labels): Testudinata, Rhyncho-cephalia, Squamata, Crocodilia, Cotylosauria, Eosuchia, Thecodontia, Saurischia, Ornithischia, Pterosauria, Ichthyosauria, Sauropterygia, Mesosauria, Protosauria, Plesiosauria, Therapsida

Subclass groupings: Anapsida, Lepidosauria, Archosauria, incertae sedis, Ichthyopterygia, Euryapsida, Synapsida

* The Tertiary is frequently divided into epochs. Beginning with the oldest, these are: Paleocene, Eocene, Oligocene, Miocene, and Pliocene. The Quaternary is also divided into Pleistocene and Recent.

FIGURE 5-3
The geologic time scale, showing the distribution of the orders of reptiles in time.

why the names of the types of skulls and of some of the suborders are derived from the Greek root *apse*, which means "arch." Thus, the diapsids are reptiles with two arches, hence two pairs of temporal openings in the skull.

RADIATION OF THE REPTILES

The reptiles of the present are but a remnant of the hordes that roamed the world during earlier geologic times. Once they had solved the problem of life away from the water, reptiles began a rapid radiation which enabled them to take advantage of all the new environments that were opened to them. They were better adapted to terrestrial conditions than were the amphibians and there were as yet no birds or mammals to dispute their possession of the land. They spread rapidly. Some took to the air. Others returned to the water, though with a curious reversal of the amphibian pattern; the aquatic reptiles either had to bear their young alive or return to land to lay their eggs. This rapid radiation of reptilian species began soon after the first appearance of the group in the Pennsylvanian, and was well under way by the end of the Paleozoic. The enormous diversity attained by the reptiles in their heyday, and the reduction they have since suffered, is shown in a recent classification, which lists more than 180 families that are entirely extinct. This is more than four times the number of families that have living members.

CLASSIFICATION OF REPTILES

Subclass Anapsida

The anapsids are those reptiles with no temporal openings in the skull. They are divided into two orders.

Order Cotylosauria. This order includes the captorhinomorphs, the primitive stem reptiles which probably arose in the early Pennsylvanian and reached a climax in the late Paleozoic. They were mostly squat, ungainly, lizardlike animals with sprawling legs and small brains.

The limbs of some of the later cotylosaurs (pareiasaurs) had rotated inward, enabling them to support the weight of a heavy body more efficiently than could the sideward sprawling limbs of the amphibians and earliest cotylosaurs. The pareiasaurs were the largest of the stem reptiles, reaching lengths of more than 300 cm. They were apparently slow-moving herbivores, perhaps something like the modern hippopotamus.

The cotylosaurs became extinct before the end of the Triassic, but they left as descendants not only the second order of anapsids—the turtles—but all later reptiles.

FIGURE 5-4
Anapsid reptiles: (*lower*) a primitive, fish-eating cotylosaur, *Limnoscelis;* (*middle*) a slow-moving, herbivorous cotylosaur, *Bradysaurus;* (*upper*) the modern Alligator Snapping Turtle, *Macrochelys.*

Order Testudinata.   Turtles, although they belong to the most primitive group of reptiles, have persisted until modern times. They have hard shells and skeletons that are heavy in proportion to their size. They frequently live in aquatic habitats where dead animals may sink and be covered with silt, thus being protected from the forces of decay and erosion while being fossilized. For these reasons the turtles have left a better fossil record than have the lizards and snakes with their more fragile skeletons and terrestrial habits.

Two suborders of the Testudinata, the Proganochelydia and Amphichelydia, are entirely extinct. When these most primitive of the turtles first appeared in the Upper Triassic, they already had the typical turtle armor fully developed. Apparently the head could not be pulled back into the shell. Most of the Amphichelydia had died out by the end of the Mesozoic, but one form, *Meiolania,* persisted in Australia until the Pleistocene. It must have been a huge beast—its skull was more than 60 cm wide.

Modern turtles are divided into two suborders: Pleurodira, the side-necked turtles, which turn the neck sideways to tuck the head under the shell; and Cryptodira, which bend the neck in a vertical sigmoid curve to withdraw the head.

Most of the families of living cryptodires were present in the Mesozoic. The Dermatemydidae mainly includes rather primitive turtles of the Jurassic, Cretaceous, and early Tertiary, but a single species, *Dermatemys mawi,* still exists in Central America. The Trionychidae, Dermochelyidae, and perhaps Testudinidae were present in the Upper Jurassic, and the Cheloniidae and Carettochelyidae in the Cretaceous. The Chelydridae have not been positively identified before the Oligocene.

There are only two families of pleurodire turtles. The Pelomedusidae go back to the Lower Cretaceous, but the Chelidae are not definitely known earlier than the Oligocene. (For descriptions of the modern families of turtles, see Chap. 15).

Subclass Lepidosauria

This is the first subclass of diapsid reptiles, those that have two temporal openings on each side. In the later lepidosaurs, one or both of the arches may be lost so the nature of the openings is obscured, but these groups can be traced to forms that were diapsid. The lepidosaurs apparently diverged from the cotylosaur stock during the Permian. Of the three orders in this subclass, two survive.

Order Eosuchia.   These were rather small reptiles of the Permian and early Mesozoic. They were lizardlike in appearance and perhaps also in behavior, but they differed from modern lizards in having a complete arch of bone that closes the lateral openings on the lower side. In true lizards this arch has

FIGURE 5-5
Lepidosaurian reptiles: (*upper left*) Tuatara, *Sphenodon*; (*upper right*) Chameleon, *Chamaeleo*; (*lower left*) Mosasaur, *Clidastes*; (*lower right*) Rock Python, *Python*.

disappeared. Most of the eosuchians had died out by mid-Triassic times. However, *Champosaurus* of the late Cretaceous and early Tertiary, a long-snouted, aquatic diapsid reptile rather like a modern gharial in build, may have been a late-surviving member of the eosuchian stock.

Order Rhynchocephalia. The beak-headed reptiles resemble the eosuchians but differ from them in having the teeth fused to the edge of the jaw rather than set in sockets, and in having an overhanging beak on the upper jaw, from which they get their name. Like the eosuchians, and unlike the Squamata, they retain the two complete temporal arches. The rhynchocephalians evolved from the eosuchians about the beginning of the Triassic. One species, *Sphenodon punctatus,* still lives on a few islands off the coast of New Zealand (see Chap. 18). This "living fossil" is among the most venerable of the vertebrates, for it appears to have survived with little change from the early Mesozoic. All other known rhynchocephalians are from the Mesozoic.

Order Squamata. This order includes the lizards and snakes, the really successful modern reptiles. They differ from the other lepidosaurs in the reduction of the bony framework on the side of the head by the loss of one or both of the arches. There are three suborders, the Lacertilia (lizards), the Serpentes (snakes), and the Amphisbaenia (ringed lizards).

The dinosaurs were so spectacular that the study of them dominated paleontology for years. Now that the major outlines of dinosaur evolution have been drawn, paleontologists are turning their attention to the lesser forms that managed somehow to survive the destruction that overwhelmed the dinosaurs and persist to the present day. Recently developed techniques of micropaleontology have stimulated the study of the tiny and fragile lizard jaws and snake vertebrae that tell the story of these little cousins of the dinosaurs. Enough has been done already to show that our limited knowledge of the fossil history of the Squamata results not from their lack of an extended evolutionary past nor their failure to leave a fossil record, but from our failure to look for the evidence. We may expect in the next few years an enormous increase in our knowledge of fossil snakes and lizards, which will undoubtedly lead to some realignment of the families. Especially we may hope that light will be thrown on the murky picture of the relationships of the snakes that are now lumped together in the family Colubridae—this is perhaps the most pressing systematic problem in herpetology.

The lacertilians probably branched off from the eosuchians some time in the Triassic. A separate infraorder, Eolacertilia, has been erected for *Kuehneosaurus* and its allies, specialized gliding lizards of the Upper Triassic. A number of forms are known from the Upper Jurassic; most of them are assigned to extinct families. Of the modern families the Agamidae, Chameleonidae, Teiidae, Scincidae, Anguidae, Varanidae, and Xenosauridae were all present in the Cretaceous. The Xantusidae are known from the Paleocene, the Gekkonidae, Iguanidae, Cordylidae, and Helodermatidae from the Eocene. The other families are unknown as fossils. (For descriptions of the modern lizard families, see Chap. 16).

The most spectacular lizards that ever lived were the mosasaurs, huge, sea-going, fish-eaters with long heads, long tails, and paddle-shaped limbs. The largest reached 12 meters in length. They were common in all seas during the Upper Cretaceous but did not survive into the Cenozoic. The amphisbaenians were present in the Paleocene.

Our knowledge of fossil snakes is even more limited than our knowledge of fossil lizards. The bones of the snake skull are loosely connected and after death are soon scattered and lost. Usually only the vertebrae are preserved in identifiable shape as fossils. The poverty of our knowledge is illustrated by the fact that out of nearly 300 genera that are included in the family Colubridae, about two dozen are known as fossils. Only for the large boids do we have anything even approaching an adequate record.

The snakes seem to have evolved from the lizards, probably toward the end of the Mesozoic, for the earliest fossil snakes are from the Cretaceous.

These include members of three extinct families, Dinilysiidae, Simolio-phiidae, and Lapparentophiidae. The modern Boidae and Aniliidae were also present in the late Cretaceous. The extinct families Palaeophiidae, Archaeo-phiidae, and Anomalophiidae and the living Typhlopidae were present in the Eocene and the Acrochordidae in the Miocene. The more advanced Colu-bridae, Elapidae, and Viperidae are also known from the Miocene. The other families are unknown as fossils. The characters that distinguish the modern families become much less definite as we trace them back to the early Cenozoic. It is probable that the radiation of the modern snakes is a recent phenomenon, and that some of the families were not present before the late Cenozoic. (For descriptions of recent families of snakes, see Chap. 17).

Subclass Archosauria

As the name implies (*archon* = ruling, *sauria* = reptiles), the members of the second subclass of diapsid reptiles were the ruling reptiles of the Mesozoic. In addition to the dinosaurs, the group includes such varied types as the crocodilians, the flying reptiles (pterosaurs), the crocodilelike phyto-saurs, and the ancestors of the modern birds. Although the end forms are very diversified, the archosaurs seem to be a natural group that may have branched off from either the primitive eosuchians or the cotylosaurs in the Permian. Although they were of little importance in the evolution of our modern reptiles, they were the lords of the earth during the Mesozoic era.

Order Thecodontia.   This Triassic order includes the primitive archo-saurian stock—the ancestors of the crocodilians, pterosaurs, dinosaurs, and also of the birds. Most of them were small, very active carnivores, rather lizard-like in build. The hind legs were always longer than the front legs, the beginning of the trend toward bipedal locomotion that was so strongly marked in many of the later archosaurs.

Also included in the Thecodontia are the specialized Phytosauria, which were common in the Upper Triassic. With long bodies, long tails, and long jaws lined with many sharp teeth, the fish-eating phytosaurs superficially resembled their contemporary relatives, the early crocodiles, but differed in one apparently vital respect: instead of having nostrils at the tip of the snout, with the nasal chamber separated from the mouth by a secondary palate, the phytosaurs had nasal openings located on top of the head between the eyes and leading directly into the mouth. Phytosaurs were less successful than the crocodiles and by Jurassic times they had disappeared.

Order Crocodilia.   The crocodilians, which appeared first in the Upper Triassic, are among the most conservative of the archosaurs. The more

FIGURE 5-6
Archosaurian reptiles: *(upper left)* a Triassic thecodont, *Saltoposuchus;* *(upper right)* a primitive pterosaur; *(center) Alligator; (lower left)* a saurischian dinosaur, *Tyrannosaurus; (lower right)* an ornithischian dinosaur, *Stegosaurus.*

specialized pterosaurs and dinosaurs, although successful for a time, died out by the end of the Mesozoic, but the crocodiles survived. Today they are the only living remnants of the once widespread and dominant archosaurian stock.

Modern crocodiles belong to two families: the Crocodylidae, which date from the Upper Cretaceous; and the Gavialidae, which were certainly present in the Miocene and may have existed as far back as the Eocene. (For descriptions of the modern families of crocodiles, see Chap. 18). In addition, fourteen extinct families are known, including the marine thalattosuchians with fishlike tail fins and paddle-shaped limbs.

Order Pterosauria. These archosaurs took to the air. They represented an evolutionary line other than the one that gave rise to the birds; their flight mechanism was more like that of the bats. They never developed feathers but depended entirely on skin membranes that stretched from the extended tips of the fourth fingers to the ankles. Their wings were long and narrow and from what we know of their habitat and manner of living it is fairly certain that they were soarers rather than strong flyers. Their small breast bones would not have permitted the attachment of muscles large enough to make them active flyers. They probably lived on cliffs near the sea coasts, feeding on fish, and depending on rising air currents for sailing and gliding. Some were only slightly larger than sparrows, but *Pteranodon*, the giant of the group, had a wingspread of about 8 meters. Pterosaurs appeared first in the Lower Jurassic and lived through the Cretaceous.

The remaining two orders of archosaurs are collectively known as the dinosaurs. They were the most impressive land animals that ever lived. Only through a knowledge of them can we grasp the full evolutionary capabilities of the reptilian type of body organization and mode of life. If it were not for the dinosaurs, we might conceivably theorize that reptiles, because of their less efficient methods of heat regulation, are unable to attain the body sizes reached by the larger mammals.

Order Saurischia.   The two orders of dinosaurs can be distinguished by the structure of the pelvic girdle. The Saurischia had a triradiate (three-branched) pelvis similar to that found in many thecodonts. The pubic bone of the Ornithischia had swung back and down to lie parallel to the ischium and had developed a broad, new, forward projection, thus forming a quadriradiate (four-branched) pelvis.

The saurischians evolved from thecodonts in the Triassic. They are divided into two groups: the carnivorous, bipedal dinosaurs (theropods), and the herbivorous, quadrupedal dinosaurs (sauropods). The theropods included a number of large, heavily built flesh-eaters, the biggest of which was the huge and spectacular king of the carnivores, *Tyrannosaurus rex*. This creature was about 15 meters long and 6 meters high. It was undoubtedly the most dangerous carnivore the world has ever known. Other theropods were small, slightly built, predaceous animals, some not much larger than a barnyard chicken.

True sauropods appeared in the Jurassic. They had reverted to quadrupedal locomotion, although their front legs were still usually much shorter than their hind legs. They were long-necked, long-tailed, and small-headed. *Brachiosaurus,* the largest, though not the longest, reached about 24 meters in length and may have weighed nearly 50 tons. Although they were the largest animals that ever roamed the earth, they were rather defenseless creatures that relied on their size and habitat for protection from the carnivorous theropods. They probably lived in marshes, grazing among the succulent plants much as cows do nowadays. The bipedal *Tyrannosaurus* would have had heavy going in such terrain.

Order Ornithischia.   The ornithischian dinosaurs differed from the saurischians in having a quadriradiate pelvis. Some were bipedal, though not as fully as most of the theropods; many lines reverted to a quadrupedal gait. They were all herbivorous. Ornithischians were late arrivals among the reptiles. They were very rare before the Upper Jurassic, but they later underwent an extensive radiation before their extinction at the close of the Cretaceous.

Such bizarre types as the armored dinosaurs, the horned dinosaurs, the aquatic duckbills, and the crested dinosaurs appeared among the ornithischians. The horned dinosaurs were rather large, though they never approached the size of the largest saurischians. In bulk and appearance an animal such as *Triceratops* must have been something like a modern rhino, with horns on the head and a bony shield projecting back from the posterior margin of the cranium over the neck and shoulder region. This was probably an efficient protection against being nipped on the back of the neck by *Tyrannosaurus*.

The armored dinosaurs were weird-looking creatures. Many had spectacular bony plates along the nape of the neck, the back, and the dorsal margin of the tail. Others had bony plates over the entire back and looked like huge tortoises. Some developed large spines and knobs along the sides of the body and on the tip of the tail, which they probably swung in defense much as a modern alligator does. A tail armed with such large bony spines was undoubtedly a very effective weapon.

The duckbill dinosaurs were aquatic. They probably fed like modern ducks, wading around the edges of marshes and reaching down to sieve out food with their large, flat, ducklike bills. Crested dinosaurs developed bony protuberances at the tops of their heads. They were cavernous structures, the internal chambers being continuous with the nasal air passages. These may have functioned as air reservoirs to permit the animals to remain submerged for long periods of time.

Subclass Uncertain

Order Mesosauria.   This order was erected for a single genus of reptiles *(Mesosaurus)* that was found in early Permian beds on either side of the South Atlantic. *Mesosaurus* had a lateral temporal opening as did the Synapsida discussed below, but the opening was placed far down on the cheek and almost certainly evolved independently. *Mesosaurus* differed greatly from the synapsids in other ways also, and it does not seem to have been related to any of the other main reptilian stocks. It may have represented an early, short-lived side branch of the Carboniferous stem reptiles. It was less than a meter in length, and had a long snout armed with many sharp teeth, a long and powerful tail, and well-developed hind legs. It probably lived in fresh water and preyed on fish.

Subclass Ichthyopterygia

The Ichthyopterygia were reptiles with a dorsal temporal opening. They were a compact group, highly modified for marine life, and showed little resemblance to any of the other reptilian stocks. The subclass contains only one order.

FIGURE 5-7
Marine reptiles: (*upper*) a plesiosaur;
(*middle*) an ichthyosaur; (*lower*) *Mesosaurus*.

Order Ichthyosauria.   The fish lizards were the dominant marine reptiles of the Mesozoic seas. They were built superficially like streamlined fish and were about the size of small porpoises. Their eyes were large and their nostrils were set just anterior to the eyes; the head joined the body with no perceptible neck. The long jaws, armed with many sharp teeth, indicate that the ichthyosaurs were fish-eaters. These animals probably occupied the same ecologic niche as do the porpoises today. They were obviously no better adapted for moving onto land than are the modern whales and it was long suspected that instead of laying eggs they bore their young alive. This was confirmed by the spectacular discovery of fossils that show traces of well-developed embryos within the body of the mother. Ichthyosaurs appeared in the early Triassic and survived until the Upper Cretaceous.

Subclass Euryapsida

These reptiles, like the ichthyosaurs, had a dorsal temporal opening, and the later forms were also marine, but their aquatic specializations were entirely different. There is little doubt that the two subclasses represent independent lines of reptilian evolution. There are three orders of Euryapsida.

Order Araeoscelidia.   This order includes some primitive, rather lizardlike animals of the Permian and Triassic. Most were slightly built little creatures with rather long, slender limbs. A grotesque member was the Triassic *Tanystropheus;* it was about 75 cm in length, and had a neck that was nearly as long as the rest of the body and tail combined. The lengthening of the neck resulted from the elongation of the individual vertebrae. When the first complete skeleton was discovered, it was found that the anterior part of the body had previously been described as belonging to a flying reptile, while the trunk had been thought to be that of a small dinosaur. No one has yet made a sensible suggestion as to its mode of life.

Order Sauropterygia.   The dominant sauropterygians were the marine plesiosaurs. Perhaps the most spectacular reptiles ever to invade the seas, they have been likened to a snake strung on the body of a turtle. The head was small, the neck long with many normal-sized vertebrae (76 in one form), rather than the elongated vertebrae characteristic of the araeoscelids. The body was short and flattened, the limbs large and heavy, the tail long. Some attained lengths of 15 meters. Presumably the plesiosaurs swam by rowing themselves through the water with their paddlelike limbs, and, like the ichthyosaurs, they were undoubtedly fish eaters. They appeared in the Middle Triassic and were common in Jurassic and Cretaceous seas.

   Also included in the Sauropterygia are the less highly specialized Triassic nothosaurs.

Order Placodontia.   The aberrant, armored placodonts of the Middle and Upper Triassic had shortened necks and broad, flat teeth for crushing molluscs.

Subclass Synapsida

Except for *Mesosaurus,* the synapsids were the only reptiles to develop lateral but not dorsal temporal openings in the skull. One of the oldest of the reptile groups, they appeared in the Carboniferous and were the commonest reptiles of the Permian. After the emergence of the dinosaurs in the Triassic, the synapsids declined rapidly, and by the close of the period had practically disappeared. But before they did so, they gave rise to the mammals, the group that replaced the dinosaurs and became the dominant animals of the Cenozoic.

The Synapsida are divided into two orders in a more or less horizontal fashion.

Order Pelycosauria.   These were the primitive synapsids of the Carboniferous and Lower Permian; a few stragglers survived into the Upper Permian and one doubtful form into the Lower Triassic. Most of them were rather small, though some of the later forms were more than 3 meters long. They all had sprawling legs like the amphibians. Some were long-snouted fish eaters, others were more terrestrial carnivores, and a few became herbivorous.

FIGURE 5-8
Synapsid reptiles: (*upper*) *Cynognathus*; (*middle*) a pelycosaur, *Dimetrodon*; (*lower*) a dicynodont, *Kannemeyeria*.

Most spectacular of the pelycosaurs were the ship lizards (e.g., *Edaphosaurus*), so named because the enormously enlongated neural spines are thought to have supported a curious, saillike fin along the midline of the back. It has been suggested that this large fin may have been a means of controlling body heat. It was probably rich in blood vessels. Spread out in the sun, it would have sped up the absorption of heat. If the animal grew overheated, it could move into a cooler, shady spot where heat would be quickly dissipated through the membrane. This mechanism, if it were truly used for thermostatic control, was not the one that finally proved successful; it remained for another group, with a different sort of thermostatic control, to lead to the mammals.

Order Therapsida.   This order of advanced synapsids appeared in the Middle Permian and its members were enormously abundant and diversified in the Upper Permian and Lower Triassic. Some were carnivorous, others herbivorous. Many therapsids were bulky animals and some attained lengths of more than 4 meters. Others were smaller, active, predaceous types. They were characterized not so much by common structural features as by the trends the various lines showed toward the mammalian condition. These trends included the development of a heterodont dentition, of a double condyle on the skull, and of a change in the position of the limbs so that they no longer sprawled sideways but were brought under the body to support it off the ground. Therapsids also developed a secondary palate, a bony plate separating the nasal chamber above from the oral chamber below. Because they do not maintain a constant body temperature and can survive with a low metabolic rate, ectothermic reptiles can tolerate fluctuation in the supply of oxygen, and may depress or even stop their breathing for a time. The endothermic mammals, in which metabolic activity continues at a steady and high rate, must breathe regularly. The development of a secondary palate permits an animal to breathe while it is holding food in its mouth. This device was a major step in evolution toward the mammals.

Since mammalian characters include physiological functions and such soft parts of the body as mammary glands and hair, it is impossible to say from fossil evidence exactly when the therapsids ceased to be reptiles and became mammals. We will leave them on the threshold and return to the modern amphibians and reptiles.

READINGS AND REFERENCES

Colbert, E. H. *The Dinosaur Book.* 2d ed. New York: McGraw-Hill, 1951. (A popular account, recent and readable.)

Gans, C., ed. *Biology of the Reptilia.* Vol. 1. New York: Academic Press, 1969.

Piveteau, Jean, ed. *Amphibiens, Reptiles, Oiseaux.* Traité de Paléontologie, vol. 5. Paris: Masson, 1955.

Romer, A. S. *Osteology of the Reptiles.* Chicago: The University of Chicago Press, 1956.

———. *Vertebrate Paleontology.* 3rd ed. Chicago: The University of Chicago Press, 1966.

———. "Early Reptilian Evolution Re-viewed." *Evolution,* vol. 21, no. 4, 1967.

Szarski, H. "The Origin of Vertebrate Foetal Membranes." *Evolution,* vol. 22, no. 1, 1968.

6

REPRODUCTION AND
LIFE HISTORY OF
AMPHIBIA

Most of the activities of an animal, whether it is feeding, seeking a safe resting place, or escaping from an enemy, are directed toward the preservation of the individual. Periodically, however, these survival activities are superseded by another set of activities that lead to the perpetuation of the race. The urge toward reproduction is so strong that the animal may change its customary way of life completely: it may stop feeding, leave its home, and travel long distances, exposing itself to enemies on the way, to breed. In this chapter we are concerned with these activities and their outcome— the development of a new generation to replace the old.

The most striking thing about amphibian reproduction is its diversity. We can hardly make a general statement about any phase of it that does not have some exceptions. Fertilization may be internal or external, the animals may lay eggs or bring forth their young alive, eggs may be laid in water or on land, there may be a larval stage or the development may be direct, parents may abandon their eggs or guard them. If we were to speak in anthropomorphic terms, we might say that the amphibians have been experimenting, trying to find the method or methods of reproduction that are best suited to life on land.

BREEDING ACTIVITIES

Breeding Season

Obviously, before an animal can breed it must be physiologically ready, that is, it must have a supply of ripe ova or sperm. Most animals are able to breed for only a relatively short period, or perhaps several short periods, during a year. An innate physiological rhythm apparently governs the development of most amphibian sperm and ova, timing it so that breeding will take place when conditions are most favorable for the development of the eggs and larvae.

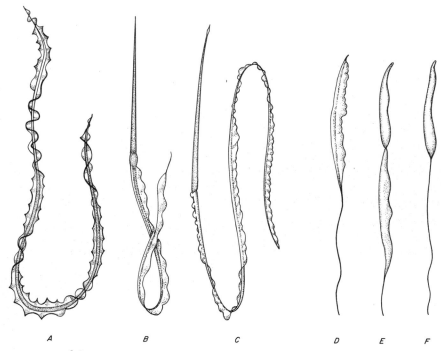

A   B   C   D   E   F

FIGURE 6-1
Spermatozoa of some amphibians: (A) *Pseudobranchus striatus*; (B) *Triturus marmoratus*; (C) *Amphiuma means*; (D) *Bombina variegatus*; (E) *Bufo vulgaris*; (F) *Hyla arborea*. [Redrawn from Noble (1954), Angel (1947), and Austin and Baker (*J. Reprod. Fertility*, 1964).]

Seasonal changes in the activity of both the anterior pituitary gland and of the gonads may be involved in the regulation of this rhythm. Males of the Common Frog of Europe *(Rana temporaria)* cannot be induced to produce sperm during the autumn and winter months. Only in the spring do the testes respond to the gonadotrophic hormones of the anterior pituitary by the formation of mature spermatozoa. On the other hand, ovulation can be induced in females of this and many other species of frogs at any time of year by injection of these hormones. That the females normally produce eggs only in the spring and summer indicates that the rhythm is controlled by cyclical activity of the pituitary. It is probable that it is also at least partly under some kind of environmental control. The activity of the anterior pituitary is known to be influenced by such external factors as light and temperature.

Once the animals are in breeding condition, breeding activity is induced by appropriate climatic factors. In the north temperate zone, where most breeding activity takes place in the spring and summer, increase in temperature probably has an important triggering effect. However, breeding is not always associated with a rise in temperature. The Florida race of the Spring Peeper *(Hyla crucifer bartramiana)* spawns during the winter months of December, January, and February, whereas the northern race *(H. c. crucifer)* breeds in April and May. Probably the temperatures at which breeding takes place are very similar for both forms. Most frogs live in the tropics where there is little seasonal change in temperature. In the "wet and dry" tropics, the onset of the rainy season is the most important triggering mechanism in inducing frogs to spawn.

In some species of frogs, the adults in one locality all reach breeding condition at about the same time; in other species they do not. If they do, then breeding is explosive, with all the reproductive activity taking place in a few days or weeks; if they do not, then breeding is protracted and may extend over a period of months. The breeding season for many frogs on Barro Colorado Island in Panama lasts eight months, the duration of the rainy season. In warm, constantly moist climates, some species may breed all year round. Spadefoot Toads *(Scaphiopus)* are explosive breeders. They lay their eggs in temporary pools that are produced by torrential downpours; most of the activity is completed twenty-four hours after the onset of such a storm. These toads cannot be said to have a true breeding season. One year they may spawn in March, the next year in September; in some years there may be several breeding periods, in other years none. The triggering mechanism is unknown; it may be either the sudden decrease in barometric pressure that accompanies these heavy storms or the saturation of the ground with water. The adults probably remain in a state of physiological readiness for most of the warmer part of the year.

There have been few reports on how many times an individual frog may breed during a season. *Rana temporaria* in England is an explosive breeder. In a study of tagged frogs in a single pool during a twelve-day breeding period, one male was seen clasping three different females on each of three successive nights. None of the marked females was observed in amplexus more than once.

A population of the Gulf Coast Toad *(Bufo valliceps)* has been reported to form several breeding congregations after heavy rains during a single season. The same males reappeared in successive choruses. Several times

FIGURE 6-2
The eggs in an opened nest of the terrestrial breeding leptodactylid *Kyarranus loveridgei* of Australia. [Photograph by John A. Moore.]

spent females that had laid eggs during an earlier breeding congregation were found in amplexus at a later one and a few females deposited two sets of eggs in one year.

Salamanders are predominantly animals of the north temperate zone and most of them breed in the spring. But there are many exceptions. In the western United States, the Del Norte Salamander *(Plethodon elongatus)* breeds in the late fall, and the Tiger Salamander *(Ambystoma t. tigrinum)* breeds in winter (January) in New Jersey and Maryland.

Breeding Sites

Different species of amphibians show a wide variety of breeding sites. This is particularly true of the frogs. They may lay eggs in open, standing water, either permanent or temporary, in running water (even mountain torrents), on the ground under stones or logs, in burrows, in mud basins constructed by the males, on the undersides of leaves suspended over water, or in water collected at the bases of the leaves of tropical air plants. Sometimes the eggs, or tadpoles, or both, are carried about by one of the parents. The young of *Rhinoderma darwini* go through their posthatching development in the vocal pouch of the male!

Salamanders show less diversity in choice of breeding sites, but even so their eggs may be laid in still water, in swift flowing streams, or on land.

Breeding Congregations

A few amphibians, including some aquatic frogs and salamanders and some terrestrial forms that lay their eggs on the ground, normally live during the nonbreeding season in their breeding places. Individuals of other species disperse widely and are found living far from the spawning sites. When the time for reproduction is at hand, they congregate in large numbers in areas much smaller than, and often of a much different character from, the non-breeding habitat. The males of both frogs and salamanders characteristically arrive before the females. These large breeding congregations increase the chances of an individual finding a mate.

The question arises of how the widely dispersed animals find their way back to the restricted breeding areas. We cannot assume that a frog remembers the route by which it traveled away from a pond when it emerged after metamorphosis a year or more earlier. Nor can we assume that it is mere aimless wandering that brings thousands of frogs of the same species together on the shores of one small lake. The physiological changes that take place when the animal reaches breeding condition apparently sensitize it to certain stimuli to which it does not respond at other times. In view of the multiplicity of breeding sites selected, the stimuli to which the animals respond must differ from species to species. Voice seems to be important in many frogs. The males nearest the breeding site begin calling first. Those farther off are guided by the calls of the earlier arrivals and hasten to join the chorus. A Southern Toad, *Bufo terrestris,* will move toward a chorus from a distance of nearly 1 km. Voice is not the only stimulus directing movement in frogs, however. *Rana temporaria* in England has been shown to migrate in response to the odor of ripening algae in the ponds. And voice is certainly not the cue for the silent salamanders, which also migrate over long distances to reach their breeding sites. Celestial cues, the odor of marsh vegetation, temperature and humidity gradients, the downhill slope of the ground, and, for aquatic salamanders, oxygen gradients and the flow of current, have all been suggested as stimuli guiding the direction of migration in amphibians.

Voice

Amphibians are the first vertebrates to have true voices. The ability to produce sounds differs in the different groups; caecilians and sirens are essentially silent as are most salamanders and female frogs. Male frogs possess some of the most remarkable voices in the animal kingdom. They have well-developed vocal cords and also large, inflatable vocal pouches to add resonance to their calls. Their habit of congregating in large groups results in choruses that can be heard for great distances and that are almost deafening when one is close. Frog calls differ in pitch, duration, frequency, and harmonics. They may consist of a single note or several. In many species two or three individuals call in sequence so that a rhythmic structure is imparted to the chorus. Each species has its own particular call which is so characteristic that a frog can be identified by its voice alone just as surely as can a bird (see Fig. 11–4). Similarity of voice between different species may be a valuable clue to relationships. Voice is such a striking feature that many frogs are named for the sounds they produce. Thus we have the Bullfrog (said to bellow like a bull), Barking Frog, Bird-voiced Treefrog, Pig Frog, Bell Frog, and Snoring Frog.

The calling of male frogs is closely associated with reproduction. In some species the voices of the earliest males to arrive at the breeding site attract other males as well as females to the same limited area. These are the species that form large choruses. An example is the well-known Spring Peeper *(Hyla crucifer)* of the eastern United States. In other species, only the females are attracted to a calling male, the breeding pairs are scattered over a wider area, and no true chorus results. This is especially true for such terrestrial forms as *Eleutherodactylus.* Sometimes several species resort to the same breeding site at the same time; it is very difficult for man to distinguish individual calls in the resulting cacophony.

Bringing the individuals together in the spawning area is not the only part played by voice during the breeding season. Each male Bullfrog *(Rana catesbeiana)* establishes a calling station, a particular spot to which he returns night after night. His voice apparently warns other males of the same species to stay away from that spot. Here calling by the frogs performs the same function as the singing of birds in the spring—that of establishing territory. In a mixed chorus where several different species are calling together, voice serves as an isolating mechanism, preventing the females from going to males of another species. Voice may also be useful in sex recognition, preventing a male from clasping another male rather than a female.

As a rule, only male frogs have well-developed vocal chords but a female may give a mercy scream when caught by a snake and some females will grunt when handled. Occasionally frogs that breed in the spring will also call in the fall. Some, such as the Squirrel Tree Frog *(Hyla squirella)* have distinctive rain calls which differ from the breeding calls and are given away from the spawning grounds in humid weather. For this reason they are often called Rain Frogs. Nonbreeding Spadefoot Toads are sometimes heard calling from underground.

At least one salamander, the Pacific Giant Salamander, *Dicamtodon ensatus,* has true vocal cords. Other salamanders and trachystomes yelp, squeak, or whistle. Many of these noises are produced accidentally but some are not. Their function is not clearly understood. They may be part of a defense mechanism, though in a few species, such as the Tiger Salamander *(Ambystoma tigrinum)* andy the European *Salamandra s. taeniata,* there may be some sexual significance.

Sex Recognition

When we speak of sex recognition there is no implication that amphibians consciously distinguish between the sexes. It is simply that the presence of a female evokes from the male a given series of reactions that are not evoked by another male. Similarly a female responds to a male as she does not to another female. We know very little of the precise mechanism of sex recognition. There are, of course, many differences between the sexes beside the primary differences in the genital systems. Some, such as the enlarged tympanum of a male Bullfrog, are permanent; others appear only with the onset of the breeding season. The latter are exemplified by the spiny nuptial pads on the forefeet of many male frogs and toads. These pads help the male grasp and hold the female, but, like the majority of secondary sex characters in amphibians, they seem to have little value in sex discrimination.

Both behavior and the odor (or some other recognizable stimulus) from the secretions of hedonic glands probably function in sex recognition by salamanders. During courtship the male prods, noses, or otherwise rubs against another animal, by his actions showing that he is a male. A ripe female accepts these advances, another male resists or runs away. Males of some European newts are brightly colored and display before the females. Sexual difference in color may aid in sex discrimination in these forms.

A female frog responds to the voice of a calling male by moving towards him. Males seem to have little ability to discriminate between the sexes but tend to grasp any object of approximately the right size that approaches. A female whose abdomen is distended with ripe eggs will be held until the eggs are deposited. A male or a female who has already laid will be quickly released. On the other hand, a frog will continue to clasp another male whose abdomen has been artificially distended. A male of the Leopard Frog *(Rana pipiens)* clasped by another male utters a little croak or chirp and is promptly released. The peculiar posture and gait assumed by a female frog during egg deposition apparently help stimulate the male to retain his grasp.

Courtship and Amplexus

Before mating takes place, many animals go through a series of actions that are designed to stimulate one or both partners to perform the sexual act. Procedures of the salamanders vary from lack of any true courtship activity in the primitive hynobiids to the quite elaborate behavior patterns—the *Liebespiel* (love play)—of the more advanced salamandrids and plethodontids.

Males of *Hynobius retardatus* arrive at breeding ponds first. The females follow a day or so later but are ignored by the males. When a female begins to lay, the male rushes forward and grabs the emerging egg sacs with his forelimbs while pushing the female away with his hind limbs. There is thus no contact and apparently no mutual stimulation between the pair until egg deposition actually begins. However, there are indications that the male *Hynobius nebulosus* selects a breeding site and the female is attracted to it, perhaps by secretions of the male cloacal glands.

The male of *Eurycea bislineata,* a plethodontid, noses the female and frequently bends his head across her cheek. She finally responds by straddling his tail and pressing her snout against the glands at its base. The tail is bent sharply aside, and the pair walks along in this fashion until mating is completed. In some species of salamanders the male creeps under the female and carries her along on his back, in others, he lashes his tail to waft the secretions of his hedonic glands toward the female. Because courtship behavior differs from species to species, it probably functions as an isolating mechanism as well as for sex recognition and stimulation.

Frogs have no Liebespiel; the calling of the male seems sufficient to stimulate the female. She indicates her readiness to spawn by approaching and allowing the male to mount her back and clasp her body with his front legs.

This is amplexus. He may clasp just behind her front legs (axillary or pectoral amplexus) or farther back, sometimes just in front of her hind legs (inguinal or pelvic amplexus). Since he faces the same way she does, he is now in a position to shed sperm over the eggs as they are extruded.

FERTILIZATION AND DEVELOPMENT

Fertilization

The object of the many complex activities discussed above is to bring together the male and female gametes so that they can fuse. This fusion of egg and sperm is fertilization, which may take place externally or internally.

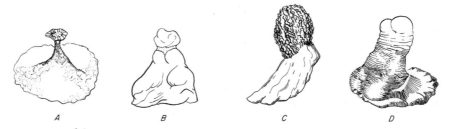

FIGURE 6-3
Salamander spermatophores: (*A*) *Notophthalmus v. viridescens*; (*B*) *Ambystoma jeffersonianum*; (*C*) *Eurycea bislineata*; (*D*) *Desmognathus f. fuscus*.
[Redrawn from Bishop, 1941.]

Fertilization among the primitive salamanders, Hynobiidae and Cryptobranchidae, is external. Among all other salamanders it is internal. During courtship the male deposits several little gelatinous packets that contain sperm (spermatophores) and the female picks them up with the lips of her cloaca. The male of the salamandrid *Euproctus* deposits the spermatophores on the female's body and uses his feet to move them to the region of the cloaca. He may even stuff them into the cloaca. *Amphiuma* is reported to transfer the spermatophore directly into the female. In the cloaca the gelatinous cap of the spermatophore dissolves and the sperm make their way to the spermatheca, a diverticulum of the roof of the cloaca. Apparently the eggs are usually fertilized as they pass by the spermatheca, but there is evidence that sperm of the European *Salamandra atra* migrate up the oviduct before the eggs are fertilized. On the other hand, the ova of the newt *Notophthalmus* cannot be fertilized until jelly layers have been deposited around them by the glands of the oviduct. Sperm high in the oviduct would be wasted. Most salamanders lay their eggs after mating but *Necturus maculosus,* which mates in the fall, does not deposit the eggs until the following spring.

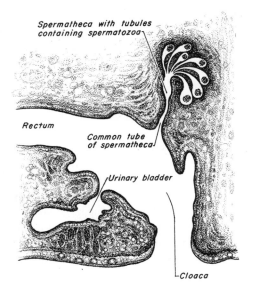

Spermatheca with tubules containing spermatozoa

Rectum

Common tube of spermatheca

Urinary bladder

Cloaca

FIGURE 6-4
The relation of the spermatheca to the cloaca in a salamander.

Frogs usually have external fertilization, with the male shedding his sperm over the eggs as they are deposited by the female in water or in some moist environment. However, the live-bearing frog of Africa, *Nectophrynoides,* practices internal fertilization, although it has no intromittent organ for the transmission of sperm. The "tail" of the American tailed frog, *Ascaphus,* is actually an extension of the cloaca that serves as an intromittent organ (see Fig. 14-3). Fertilization is internal in these mountain torrent forms; this is perhaps to prevent the sperm from being swept downstream before the eggs are fertilized.

The male caecilian has an intromittent organ that injects sperm into the cloaca of the female. Some caecilians are live bearers. Internal fertilization must be the rule in this group but we know almost nothing of how the male finds the female or where and in what manner mating takes place. Nothing is known about mating among the trachystomes.

Eggs

Amphibian eggs are generally mesolecithal. But the amount of yolk varies widely within the groups. Species that lay their eggs on land and undergo direct development have larger yolks than do species that lay in water. Thus, the terrestrial egg of the Cliff Frog *(Syrrhophus marnocki)* has a yolk 4 mm in diameter, whereas the yolk of the aquatic eggs of the Cricket Frog *(Acris gryllus)* is only 1 mm in diameter, although both adults are about the same size.

There is enormous diversity in the way eggs are deposited by different species of amphibians. Generally, however, glands in the walls of the oviduct secrete a gelatinous substance around them as they pass down to the cloaca. This substance swells in water and forms a protective jelly. Sometimes many eggs are enclosed in a single jelly mass, sometimes the eggs are separate. The jelly may be in the shape of long strings, in one large, irregular packet, or in several smaller packets deposited separately by the female. Eggs may float on the surface of the water, sink to the bottom, be attached to the stems of aquatic vegetation, or to the undersides of leaves, sticks, or rocks in the water; they may also be laid on land.

The numbers of eggs laid varies greatly from species to species. The tiny *Sminthillus* of Cuba is said to lay only one egg. On the other hand, some ranas and bufos lay more than twenty thousand in one spawning season. The number of eggs depends in part on the size of the animal; there is simply room for more eggs to develop in the body of a female of a large species than in that of a small one. But it also depends on whether the eggs are laid in the open or in some protected spot and whether they are guarded by one of the parents. Species of *Eleutherodactylus,* which lay under rocks and logs on the ground and then guard the eggs, have clutches of less than a hundred. The bufos, which spawn in open water and abandon the eggs as soon as laid, have some of the largest clutches reported for any frog.

In general, salamanders lay fewer eggs than frogs, but some may deposit as many as several hundred. Internal fertilization insures that fewer eggs will be unfertilized and hence be wasted, and the salamanders are more likely to guard their clutches.

Developmental History

Most amphibians differ from the other tetrapods in having two definitive developmental periods: embryonic development prior to hatching, and post-hatching development until metamophosis to the adult form. Young in the posthatching stage are known as larvae among the salamanders and as tad-

---

FIGURE 6-5   *(See facing page)*

The development of *Necturus maculosus. A,* Side view of egg 1 day and 8 hours after deposition, showing second and third cleavage grooves. *B,* Bottom view of egg 6 days and 16 hours old. The crescentic blastopore lip sharply separates the large yolk cells from the small cells of the blastodisc. *C,* Bottom view of egg 10 days and 10 hours old, showing large circular blastopore. *D,* Top view of egg 14 days and 4 hours old. Blastopore smaller. The beginning of neural fold formation, especially anteriorly. *E,* Top view of egg 15 days and 15 hours old. Yolk plug still visible. Neural fold prominent. Its free ends reach nearly to the blastopore. *F,* Top view of egg 18 days and 15 hours old with three or four pairs of body segments visible. *G,* Dorsolateral view of embryo 22 days and 17 hours old; length 8 mm; 16 to 18 body segments. *H,* Side view of embryo 26 days old; length 11 mm; 26 to 27 body segments; eye, ear, nasal pits, and mouth well defined. *I,* Side view of embryo 36 days and 16 hours old; length 16 mm; 36 to 38 body segments. *J,* Side view of larva 49 days old; length 21 mm. *K,* Side view of larva 97 days old; length 34 mm. [From Noble, *The Biology of the Amphibia,* Dover Publications, 1954.]

poles among the anurans. Occasionally there is no larval period and the young hatch (or are born) as miniature replicas of the adults.

Oviparity, Ovoviviparity, and Viviparity.   Usually the young of amphibians hatch after the eggs have been laid and the animals are said to be oviparous. In rare instances, however, the eggs are retained in the body of the female while they pass through their embryonic development and the young are "born alive." If the developing embryo in the mother's body is nourished

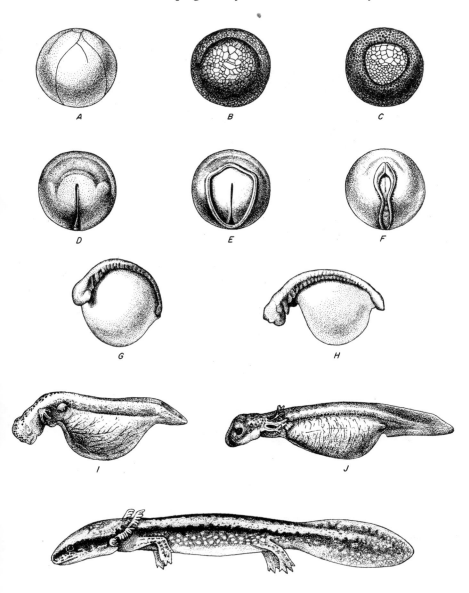

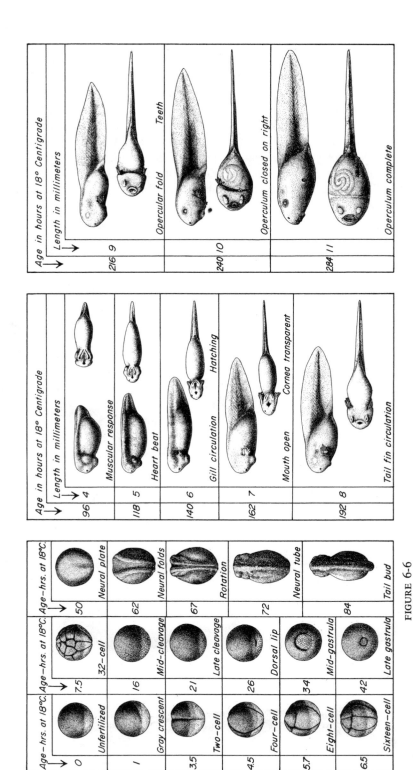

FIGURE 6-6

The external morphological features of the developing embryo of the frog *Rana pipiens*.
[Redrawn from Shumway, 1940.]

entirely by food stored in the yolk of the egg, the animal is ovoviviparous. If the embryo obtains part of its food from maternal tissues (the walls of the Müllerian duct), the animal is viviparous. The European *Salamandra* is ovoviviparous and bears living young. The larvae of one species *(S. atra)* metamorphose before birth. The African frog *Nectophrynoides* is also ovoviviparous and the young metamorphose before they are born. Some caecilians exhibit a specialized form of viviparity: the young are nourished by oil glands in the lining of the oviduct and, like those of *Nectophrynoides,* resemble adults when born. Other caecilians, however, are oviparous.

Embryonic Development.   The pattern of early development of the amphibians is typical for vertebrates that have mesolecithal eggs. The stages are: early cleavage, blastula formation, gastrulation, neurulation, tail-bud stage, the period of organogeny (formation of organs), the period of early heartbeat and development of gill buds, and finally the development of gill circulation and hatching. The details of these stages can be found in any text of embryology and need not be discussed here. Figure 6-5 shows the external morphology of a developing salamander, *Necturus;* hatching takes place about at stage J. Table 6-1 lists comparable stages in the development of the Leopard Frog, *Rana pipiens.* As the table shows, there is some variation in developmental rate, which is determined, at least in part, by temperature. Hatching takes place at about the time the gill circulation develops, or at about six days at 18°C (Fig. 6-6).

Posthatching Development.   The degree of development reached at hatching varies greatly in the amphibians. Some frogs and salamanders pass through metamorphosis in the egg or hatch in an advanced condition so that they are nearly ready to metamorphose. Most frogs hatch as tadpoles, which are quite different in appearance from their parents, and must go through a continuing period of development before they are ready to metamorphose.

   The early tadpole of *Rana pipiens* is a small, poorly developed animal, seemingly ill-equipped to meet the vicissitudes of life outside the protecting egg. Its mouth has not yet broken through, its eyes are not formed, its gills not fully developed. For a few days it remains attached to some object by a pair of V-shaped suckers below the indentation where its mouth will be and draws nourishment from its remaining supply of yolk. A small, round mouth armed with horny teeth and jaws appears, the eyes take shape, the gills grow rapidly, and the tail elongates. The tadpole is soon ready to take up a free-swimming, food-finding existence, but it still does not look like a frog. An operculum grows posteriorly to cover the external gills, which are replaced by a new set of gills. Gradually the legs develop. The hind legs emerge as tiny buds, and later develop toes and joints. The front legs grow at the same time but are hidden by the operculum. When the front legs do

TABLE 6-1
Development Stages of *Rana pipiens*

Stage	At 18°C	At 25°C
Fertilization	0 hours	0 hours
Gray crescent	1 hour	½–1 hour
Rotation	1½ hours	1 hour
Two cells	3½ hours	2½ hours
Four cells	4½ hours	3½ hours
Eight cells	5½ hours	4½ hours
Blastula	18 hours	12 hours
Gastrula	34 hours	20 hours
Yolk plug	42 hours	32 hours
Neural plate	50 hours	40 hours
Neural folds	62 hours	48 hours
Ciliary movement	67 hours	52 hours
Neural tube	72 hours	56 hours
Tail bud	84 hours	66 hours
Muscular movement	96 hours	76 hours
Heart beat	5 days	4 days
Gill circulation	6 days	5 days
Tail fin circulation	8 days	6½ days
Development of operculum	9 days	7½ days
Operculum complete	12 days	10 days
Metamorphosis	3 months	2½ months

Source: Rugh, 1951.

emerge, the left goes through the spiracle, the right through the opercular wall. Shortly before metamorphosis the tadpole comes frequently to the surface to gulp air. Its lungs are then about to replace the gills as the organs which supply oxygen to the blood. The long, coiled intestine that is necessitated by the vegetarian diet of the tadpole shortens because the adult frog is a flesh eater. The gills degenerate, the horny jaws and teeth are shed along with the tadpole skin, the round mouth widens, the tail is resorbed (although a vestige still remains), and the little animal, now a frog, hops onto land.

Salamander larvae are quite different from tadpoles. The frog tadpole has one or a pair of suckers on the ventral side of the head that persist from before to shortly after hatching; these are lacking in salamander larvae. On the other hand, the larvae of many salamander species have balancers, long, rod-like structures, one on each side, that protrude from the sides of the head below the eyes. The salamander larva has three pairs of gills, the frog tadpole only two well-developed pairs, the upper or third pair being rudimentary. After hatching the tadpole develops an operculum that covers the gills (gills are usually evident in salamander larvae until the time of meta-

SALAMANDER LARVA   FROG TADPOLE

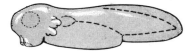

1. SHORTLY BEFORE HATCHING

Long body
No adhesive organs
Three pairs of gill rudiments

Shorter body
Prominent adhesive organs
Only two pairs of gill rudiments distinct

2. SHORTLY AFTER HATCHING

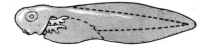

Long body
No adhesive organs
Balancers present in some species
Mouth becomes large
Three pairs of well developed external gills,
* the uppermost typically the longest*
Opercular folds simple

Short body
Prominent adhesive organs
No balancers
Mouth remains small
Only two pairs of well developed external gills,
* third pair (uppermost) rudimentary*
Opercular folds develop rapidly, soon cover gills

3. LARVAL CHARACTERS FULLY DEVELOPED

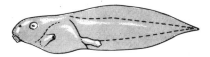

Long body
Mouth large, with simple labial folds
External gills present
Opercular folds open and loosely overhang
* sides of neck; no spiracle*
Foreleg buds external, clearly visible
* throughout their development*

Short, egg-shaped head and body unit
Mouth small, with complex accessory parts
No external gills
Operculum closed except for a small spiracle
* draining gill chamber*
Foreleg buds internal, concealed under closed
* operculum until metamorphosis*

FIGURE 6-7
Comparison of salamander larvae and frog tadpoles. [After Orton, *Turtox News*, 1950.]

morphosis). The legs of salamander larvae are often well developed at hatching. The mouth is like that of the adult. Thus, the mature salamander larva resembles the adult much more closely than the tadpole resembles the frog.

Larval Types.   Differences in posthatching habitat are correlated with morphological differences in the developing larvae. Salamanders have three basic types of life histories: some are specialized toward breeding in open bodies of quiet water; some are modified for a mountain stream existence; some have become specialized for terrestrial life.

Warm, still pond water has relatively little oxygen dissolved in it; therefore a pond-type larva needs large respiratory surfaces to absorb enough

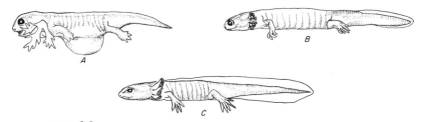

FIGURE 6-8
The principal types of salamander larvae: (*A*) terrestrial type, *Plethodon vandykei;* (*B*) mountain brook type, *Dicamptodon ensatus*; (*C*) pond type, *Ambystoma gracile.* [From Noble, *The Biology of the Amphibia*, Dover Publications, 1954.]

oxygen for its needs. For example, the pond-type larva of *Ambystoma* has long, very filamentous gills, and a well-developed tail fin that extends over its back to form a dorsal body fin.

In the cool, rapidly flowing water of mountain torrents, plenty of oxygen is available. The problem faced by larvae in this environment is rather to keep from being swept downstream by the current. The brook-type larva of *Dicamptodon ensatus* is streamlined, with a muscular tail, no dorsal body fin, a reduced tail fin, and small gills.

Finally, the terrestrial salamanders, such as *Plethodon,* have a tiny, short-bodied larva with gills that lack filaments; these gills reduce before hatching. The larva has a rather well-developed yolk mass in the gut and a short tail with no fin. These tiny creatures hatch from eggs laid on land and soon become entirely terrestrial salamanders without ever going through an aquatic larval stage.

Because tadpoles are found in a wider variety of habitats than salamander larvae, they also show more morphological diversity. Here again we find a pond type—the typical, deep-bodied, high-finned pollywog. Many mountain brook tadpoles are streamlined and have strong tails with reduced fins. The tadpole of the genus *Staurois* from the mountains of southeastern Asia has an adhesive disc on the ventral surface by which it clings to rocks and so avoids being swept away by the current. Other mountain brook tadpoles have large suctorial lips. Some torrent tadpoles have flattened bodies that are adapted for gliding over wet stones.

Some species of tree frogs *(Hyla)* lay their eggs in the tiny pools of water that are held in the axils of the leaves of bromeliads. The tadpoles are slim, with long, whiplike tails and reduced gills. Apparently they get most of their oxygen directly from the atmosphere. They feed on frog eggs, either of their own or other species, and probably also on other tadpoles. Their jaws are strong, but the larval tooth rows have been reduced. Tadpoles of the African *Hoplophryne* develop in similar habitats and show similar modifications.

Embryos of frogs that lack a free-living larval stage may yet show most of the typical tadpole structures, such as the operculum and the larval teeth and jaws. In others these structures are reduced and tend to disappear. The developing young of *Rhinoderma darwini* are distinctly tadpolelike, with closed operculum, spiracle, coiled intestine, and larval mouth parts, although the teeth and jaws never harden.

Duration of Larval Stages.   The length of the larval period varies widely, not only from species to species, but also within a given species. As with the eggs, higher temperatures accelerate the rate of development. Factors other than temperature are known to affect the developmental rate of *Scaphiopus*. The eggs are laid in temporary pools and development is very rapid since the toads must metamorphose before the pools dry up. When some tadpoles from a drying pond were transferred to an aquarium containing pond water at the same temperature, they remained tadpoles for nearly a month after those from the pond had emerged. Overcrowding, a heavy concentration of chemicals in the water, an increase in the available food supply, probably all accelerate metamorphosis.

Spadefoot tadpoles have been known to metamorphose as little as 12 days after egg deposition. Most amphibians probably remain from a minimum of a few months to a maximum of a year in the embryonic and larval stages. However, the Bullfrog *(Rana catesbeiana)* usually spends 14 to 16 months as a tadpole and may not metamorphose until after its third winter.

Recent evidence suggests that the duration of the tadpole stage in frogs is correlated with the amount of DNA per nucleus. The bullfrog has about eight times as much DNA as has *Scaphiopus holbrooki*. The frogs with the lowest DNA/nucleus values *(Scaphiopus, Hyla septentrionalis, Pseudis paradoxa)* all breed in temporary waters. The frogs with the highest values (members of the genus *Rana*) breed in permanent waters. The low levels of nuclear DNA and rapid development rates found in some frogs may explain why frogs have been more successful than salamanders in invading arid regions where development must take place in temporary waters.

Paedogenesis and Neoteny

Trachystomes, proteid salamanders, and some plethodontids never complete metamorphosis. They retain their gills and other larval structures even after they become sexually mature. This is *paedogenesis*: a genetically fixed condition in which the tissues fail to respond to the secretions of the thyroid gland that bring about metamorphosis in other forms.

Some salamanders, including many ambystomatids, fail to metamorphose when environmental conditions, such as low temperatures or lack of iodine in the water, inhibit the action of the thyroid. They become sexually mature and reproduce in the larval state. In contrast to the paedogenic forms, they are still capable of metamorphosing if the environmental conditions are changed. When such animals are fed thyroid extract, or are simply transferred to water with a higher iodine content, they soon metamorphose. Failure to metamorphose because of environmental conditions is called *neoteny.*

GROWTH AND LONGEVITY

Even after metamorphosis, amphibians must go through a period of growth and development before they become sexually mature. Some species reach breeding age in about a year. Others, particularly the larger forms like the Bullfrog *(Rana catesbeiana),* are not ready to breed until two or three years after metamorphosis, and the Mudpuppy *(Necturus maculosus),* not for several years after hatching.

Growth does not stop when an amphibian reaches sexual maturity. Many, perhaps all, continue to grow, although very slowly, for the rest of their lives.

Maximum age is very difficult to determine for animals living under natural conditions; it is much easier with animals reared in captivity. But such

TABLE 6-2

Longevity of Some Species of Amphibians

Order, family, and species	Maximum age (years)
Trachystomata	
Sirenidae *(Siren lacertina)*	25
Caudata	
Cryptobranchidae *(Andrias japonicus)*	55
Ambystomatidae *(Ambystoma maculatum)*	25
Salamandridae *(Triturus pyrrhogaster)*	25
Amphiumidae *(Amphiuma means)*	27
Proteidae *(Proteus anguinus)*	15
Anura	
Pipidae *(Xenopus laevis)*	15
Discoglossidae *(Bombina bombina)*	20
Pelobatidae *(Pelobates fuscus)*	11
Leptodactylidae *(Leptodactylus pentadactylus)*	12
Bufonidae *(Bufo bufo)*	36
Hylidae *(Hyla caerulea)*	16
Ranidae *(Rana catesbeiana)*	16
Rhacophoridae *(Kassina weali)*	9
Microhylidae *(Kaloula pulchra)*	6
(Gastrophryne carolinensis)	6

sheltered forms probably survive much longer than wild animals. The figures in Table 6-2 are for captive specimens and do not indicate longevity under natural conditions. Although the table is incomplete it does indicate that some of these small animals are capable of living for surprisingly long times under favorable conditions. It also shows that the sluggish salamanders apparently live longer than do the more lively frogs.

We know nothing of the reproductive habits of the early amphibians but it is doubtful that all the diversity shown by the present-day forms is of recent origin. The ancestral groups must at least have had considerable genetic plasticity to have allowed such diversity to develop. Most of the early forms failed to find a really satisfactory solution to the problem of reproduction on land and their descendants have remained amphibians. But members of one group, at least, did evolve a practical method of terrestrial reproduction by means of the amniote egg, and they became the reptiles.

READINGS AND REFERENCES

Angel, F. *Vie et Moeurs des Amphibiens*, Paris: Payot, 1947. (A semipopular account of the habits of amphibians, cosmopolitan in scope.)

Bishop, S. C. "The Salamanders of New York." *New York State Museum Bulletin*, no. 324, 1941.

Cochran, D. M. *Living Amphibians of the World*. New York: Hanover House, 1961. (A beautifully and profusely illustrated account of the amphibians, with much information on life histories.)

Mertens, R. *The World of Amphibians and Reptiles*. Translated by H. W. Parker. New York: McGraw-Hill, 1960. (A fascinating, splendidly illustrated book, written by one world-famous herpetologist and translated by another.)

Noble, G. K. *Biology of the Amphibia*. New York: McGraw-Hill, 1931. Reprinted by Dover Publications, 1954.

Oliver, J. A. *North American Amphibians and Reptiles*. Princeton, N.J.: Van Nostrand, 1955. (A good, recent natural history of the herptiles of the United States and Canada.)

Rugh, R. *The Frog: Its Reproduction and Development*. Philadelphia: Blakiston, 1951.

Salthe, S. N. "Courtship Patterns and the Phylogeny of the Urodeles." *Copeia*, vol. 1967, no. 1, 1967.

Shumway, W. "Stages in the Normal Development of *Rana pipiens*." *The Anatomical Record*, vol. 78, no. 2, 1940.

Smith, M. A. *The British Amphibians and Reptiles*. Rev. ed. London: Collins, 1954. (The best natural history accounts available, though limited to the fauna of the British Isles.)

7

REPRODUCTION AND LIFE HISTORY OF REPTILIA

When a group of animals makes a major shift from one mode of existence to another, there must be a close correlation in evolutionary changes of structure, function, and activity. Nowhere is this more marvelously shown than in the change in the reproductive pattern that allowed the reptiles to free themselves completely from the need to return to the water to breed. Terrestrial breeding habits, internal fertilization, shell, and extraembryonic membranes go hand in hand to produce the truly terrestrial egg of the reptile.

Sperm can travel only in a fluid medium. When fertilization takes place away from water, the fluid carrying the sperm must be protected from desiccation. The best method for insuring this is emission by the male of seminal fluid into the reproductive tract of the female: the act of copulation. Since the ovum is fertilized before it leaves the body of the female, it can be provided with a protective shell by glands of the oviduct.

Caecilians, most salamanders, and a few frogs evolved internal fertilization, but they never developed a shelled, amniote egg to go with it and so have been unable to exploit to the full the possibilities of this more efficient method of fertilization.

DIFFICULTIES IN THE STUDY
OF REPTILIAN REPRODUCTION

Reproduction of reptiles is more difficult to study than is that of most amphibians. Reptiles do not, as a rule, come together in large breeding aggregations, nor do they advertise their intentions as do the vociferous frogs. It is thus far more difficult for the herpetologist to be in the right place at the right time to observe reptilian mating. Information gathered from zoo specimens is helpful but is always questionable because the conditions of captivity may have modified the behavior of the animals.

Two other phenomena that complicate the study of reproduction in the reptiles are delayed fertilization and the initiation of development before egg deposition.

Delayed Fertilization

Among most vertebrates, fertilization takes place shortly after the emission of the sperm by the male, and indeed the sperm lose their ability to activate the ova and initiate development after a few hours, or a few days at most. In many reptiles and a few salamanders this is not so. Sperm may be stored in an inactive state in the female reproductive tract and may retain their ability to fertilize the ova for months or even years.

TABLE 7-1

Viability of Sperm of *Malaclemys*
after Separation of Males and Females

Length of separation (years)	Number of eggs laid	Number fertile
1	124	123
2	116	102
3	130	39
4	108	4

Table 7-1 shows the length of time ten female Diamondback Terrapins *(Malaclemys)* had been separated from males, the number of eggs laid each year, and the number of fertile eggs. There was a decided decline in the number of fertile eggs after two years, but a few of the sperm were viable at four years. Female Box Turtles *(Terrapene carolina)* have also laid fertile eggs two, three, and four years after separation from males.

Although delayed fertilization is well known in the turtles, they have not been shown to have special storage structures or seminal receptacles in association with the Müllerian ducts of the female. In the Squamata however, such structures are known in both the lizards and the snakes.

The seminal receptacles of the lizards are small tubules in the vaginal portion of the Müllerian ducts. During the breeding season of *Anolis carolinensis* sperm may be found in the lumen of the oviduct as well as in the receptacles but during the nonbreeding season the sperm are confined mainly to the receptacles. It seems probable that sperm survival is prolonged in the tubules: it is known that viable sperm persist in female *Anolis* for at least seven months, six months in the chameleon, *Microsaura pumila,* nearly four months in the gecko, *Eublepharis macularis,* and nearly three months in *Uta stansburiana.*

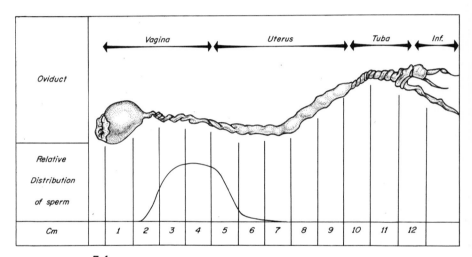

FIGURE 7-1

Diagram illustrating the relative concentration of sperm in the various regions of the "ripe" female oviduct of the Prairie Rattlesnake, *Crotalus v. viridis.* [After Ludwig and Rahn, *Copeia,* 1943.]

Snakes have seminal receptacles but instead of being in the vaginal region of the oviduct they are located at the base of the infundibulum (the funnel-shaped part leading from the ostium). As in the lizards sperm may be found in both the receptacles and the lumen of the oviduct. Figure 7-1 shows the distribution of luminal sperm in the oviduct of the Prairie Rattlesnake, *Crotalus v. viridis,* shortly before ovulation. Table 7-2 shows how long sperm remain viable in the oviducts of several species of snakes after separation from males.

Delayed fertilization or the presence of seminal receptacles have not been reported in the crocodilians or in *Sphenodon.*

TABLE 7-2

Viability of Sperm in Female Snakes
after Separation from Males

Species	Duration of viability (months)
Causas rhombeatus	5
Drymarchon corais	52
Leptodeira annulata polysticta	60
Storeria dekayi	3
Thamnophis s. sirtalis	3

Oviparity, Ovoviviparity, and Viviparity

Most reptiles are oviparous. Some lizards and snakes are ovoviviparous, with the eggs hatching either in the oviduct or just after they are laid. There is really no sharp distinction between oviparity and ovoviviparity, since many oviparous forms lay eggs in which the embryos have begun to develop. Eggs of some colubrid snakes, examined at the time of deposition, contained embryos that ranged from 15 to 55 mm in length. Such eggs may not hatch for two or three months. Alligator eggs undergo some development in the oviduct before deposition. This is also true for some of the spiny lizards *(Sceloporus)*.

Of two species in a single genus, one may be oviparous, the other ovoviviparous. European species of snakes of the genus *Natrix* lay eggs; North American species bear living young. The condition may not even be fixed in a single species. The European lizard, *Lacerta vivipara*, which bears living young throughout most of its range, lays eggs in the Pyrenees.

True viviparity implies exchange of materials between the embryonic and maternal bloodstreams. This condition is approached by a few of the Squamata. Lizards of the family Xantusidae form a shell around the egg in the oviduct but it soon disintegrates and a kind of primitive placenta develops. This provides nourishment for the embryo, which nearly doubles in weight between the time the shell disappears and the time of birth. Similar placenta formation has been observed in the European skink, *Chalcides ocellatus*, in Australia skinks of the genera *Tiliqua* and *Lygosoma*, and in the Australian snake, *Denisonia*. Histological studies of the uterine portion of the oviduct of *Denisonia* show that there is a partial degeneration of the maternal epithelium, permitting diffusion between the bloodstreams of the mother and the embryo.

The oviducts of lizards and snakes lack well-defined albuminous glands in the upper portions. The albumen in a bird's egg contributes to the amniotic fluid, which protects the embryo from desiccation. Terrestrial eggs that are not provided with a sufficient supply of albumen must be laid in surroundings moist enough to allow water to enter through the permeable shell. Such eggs swell noticeably after deposition. (Even eggs that have albumen are permeable and may lose too much water and desiccate in an arid environment.) It may be that the tendency of lizards and snakes to retain the eggs in the constantly moist environment of the oviduct for varying lengths of time, thereby leading to ovoviviparity and even viviparity, is a compensatory mechanism for the lack of albumen. Such a mechanism may have been a step in the development of the mammals from the reptiles. It has also been suggested that the retention of the eggs in the oviduct by an animal that had developed the ability to regulate its body temperature was a protective device against exposure of the eggs to low temperatures and that this may have been a major factor in the evolution of the viviparous mammals from the oviparous reptiles. Snakes and lizards that live at high altitudes or latitudes typically bear living young.

Amplexus, egg deposition, fertilization, and the beginning of embryonic development follow one another so closely in most frogs that a single series of observations may give us a fairly clear outline of the entire reproductive pattern of a species. The pattern of reptilian reproduction is not so apparent. Copulation may take place in the fall, but egg deposition may not occur until the following spring. Even when eggs are laid shortly after mating, it is still possible that the ova were fertilized by sperm received in a previous copulation. Because the initiation of development does not necessarily coincide with the time either of copulation or of egg deposition, it is not possible to calculate precisely the length of the embryonic period. Long, patient hours in the field, the dissection of many specimens in the laboratory, and much luck are necessary for the study of the life history of a reptile.

BREEDING HABITS

Sexual Encounters

One of the advantages of terrestrial reproduction over the aquatic reproduction of most amphibians is that the egg clutches can be more widely dispersed. When all the females of a population deposit their eggs in a few places, a single catastrophe, such as the too-rapid drying of temporary ponds, can bring to naught the reproductive activity of a whole season. Heavy concentrations of developing young bring heavy concentrations of predators, and competition between the young for available food may be very intense. Reptiles thus do not need to lay as many eggs as frogs to insure survival of enough young to maintain the population.

But terrestrial reproduction poses a different problem. Except for the sea turtles, most reptiles do not come together in restricted areas to form large breeding aggregations. A male must find receptive females in a population that may be widely dispersed. Because of this, the advantages of delayed fertilization are clear. If a female *Malaclemys* has mated one year, her ability to lay fertile eggs the following year does not depend on her being found again by a male. Moreover, copulation need not take place only at the time of ovulation. Some species of lizards and snakes that normally mate in the spring may also copulate in the fall, with deposition of fertile eggs the following spring. Two captive females of the European Ringed Snake *(Natrix natrix helvetica),* normally a spring breeder, mated in September and laid fertile eggs the following March.

Many northern snakes hibernate in communal dens. One such den, in an ant hill, contained 148 Smooth Green Snakes *(Opheodrys vernalis),* 101 Red-bellied Snakes *(Storeria occipitomaculata),* and 8 Great Plains Garter Snakes *(Thamnophis radix).* When snakes emerge from hibernation in the spring, they tend to remain for a while in the vicinity of the den. Since numbers of both sexes are thus present in the same restricted area, it is easy for males to find females, and most mating takes place at that time. In warm climates where snakes do not hibernate, males probably find the females by wandering.

Many lizards live in loosely knit colonies where the chances of the two sexes encountering each other are good. The circumscribed habitat of the freshwater turtles also increases the chances of the male finding a female. Males of the terrestrial tortoises seem to wander more during the breeding season than at other times.

Female marine turtles gather in large numbers at certain beaches to lay their eggs. The males congregate just off shore and mate with the females as they leave the beaches. Thus the sperm to fertilize next year's eggs are received at the time the current year's eggs are laid.

Sexual Dimorphism

The degree of sexual dimorphism varies in different orders of reptiles. Among the turtles, the plastron of a sexually mature male may be more concave beneath than that of the female. This undoubtedly helps him retain his balance on the convex carapace of his mate during copulation. Tails may also vary in length, those of males usually being longer. In some species the male has long claws on his forefeet with which he tickles the cheeks of the female during courtship. Females are often much larger than males, but in some species males are larger.

Morphological differences between male and female lizards are often quite apparent. The male is apt to be larger and may have better developed dorsal or caudal crests. Males of some species display brilliantly colored throat fans during the breeding season. An example is the common Anole

(Anolis carolinensis) of the southeastern United States; a bright green male spreading a bright red throat fan is a spectacular sight. Sometimes there are differences in body color. Males of *Sceloporus undulatus* have brighter blue patches on the sides than do females. Male lizards may also have femoral or preanal pores and modified scales in the region of the vent. In some chameleons *(Chamaeleo),* horns or dermal protuberances near the end of the snout or in front of the eyes are better developed in males than in females.

Sexual dimorphism is usually much less evident in snakes. Sometimes no distinguishing characters are visible. The main differences are in the number and form of the scales on the ventral side of the body and in the length of the tail. Males of the Asp-viper *(Vipera aspis)* have from 134 to 158 abdominal scales and from 32 to 49 subcaudals (scales on the underside of the tail). Females have 141 to 169 abdominals and 30 to 42 subcaudals. The body of the male is thus relatively shorter than that of the female, and his tail is relatively longer. The base of the tail in the male is somewhat more dilated to accommodate the hemipenes, and the scales are often more heavily keeled, as in *Homalopsis buccata,* an aquatic Asiatic snake. Males of *Chironius carinatus* of Central and South America have the median rows of vertebral scales strongly keeled whereas those of the female are only slightly, if at all, keeled. Among the Boidae, the vestigial hind limbs, or spurs, are better developed in the males than in the females, some of which lack them entirely. Tubercles formed of dermal papillae are scattered over the chin or the sides of the face of the males of a few species of snakes, such as *Natrix rhombifera* of North America and *Aspidura trachyprocta* of Ceylon. One sex or the other may be more brightly marked, especially among the vipers. There is no general rule for determining the sex of a snake by external examination. Each species must be studied separately.

Sex Recognition

In general, the female reptile plays a relatively passive role in sexual encounters. It is the male that actively pursues and courts, and hence must be able to distinguish between the sexes. There is evidence that the male of many reptiles distinguishes between "other male" and "not male" rather than recognizing a female as such. That is, another animal, or even an inanimate object, not recognized as a male, is regarded as female. The nocturnal Banded Gecko *(Coleonyx variegatus)* depends primarily on behavior as a means of sex recognition. Anesthetized males were treated as females by other males. When the blue patches on the sides of males of the diurnal Fence Lizard *(Sceloporus undulatus)* were painted out, they were treated as females by males. Males of marine turtles are sometimes lured into the nets of fishermen by wooden decoys shaped roughly like females.

Appearance, behavior, and odor play a part in the sex recognition of the reptiles. Head bobbing—a characteristic gesture that has been observed in many lizards, snakes, and turtles—seems to function in sex recognition. Chemical stimuli are important among snakes. A male snake will follow the trail left by drawing the skin of a female across a plate but will not follow one made by smears from her cloacal glands. Among diurnal lizards with marked sexual dimorphism, sight probably plays the major role but odor may also be important. We have watched a male Broad-headed Skink *(Eumeces laticeps)* trail a female across the ground, though she was hidden from sight in a hollow tree.

Some male lizards establish a territory that they defend against other males. During the breeding season a male Anole spreads his throat fan and bobs up and down at sight of an intruder. If the latter is another male, it either retreats or displays its fan in return. If both continue to display, a fight may ensue; the loser is driven from the area. A female has no throat fan to display and it is her failure to respond as a male that apparently results in her being recognized as a female; she is then either ignored or courted.

The roaring of a bull alligator, which was once a commonplace springtime sound of the southern marshes and is now too seldom heard, may serve both to warn away other males and to attract females.

The combat dance of male snakes may also play a part in sex recognition. H. T. J. Prior *(Country-Side,* vol. 9, p. 492, 1933) gives a vivid account of the dance of a pair of vipers *(Vipera berus):*

> In May I was witness to a peculiar dance, the participants of which were two male adders, which took place in a shallow leaf-carpeted ditch. I was drawn to the spot by the rustling of dead leaves and saw the silver-grey, black-blotched bodies of two male adders writhing and coiling about each other in a fantastic 'dance.' Both would sway sideways in contrary directions, then slowly sweep back, the bodies meeting and then crossing so that each ultimately reached the position occupied by the other.
>
> After a series of these slow movements, the contestants became suddenly excited, darting and ducking until one was in a position to force his opponent to the ground. While these rapid head movements were being executed the snakes' bodies were coiled in a succession of undulating patterns. After regaining his breath the fallen viper would erect himself again, sidle up to his conqueror and incite him to further strife. I watched the monotonously reiterated posturings until the smaller of the two snakes freed himself and made off with amazing celerity, closely pursued by the other until out of sight. The female, the cause of this, was lying coiled close at hand.

Similar dances during the mating season have been observed in a number of other species. Some keep their bodies in a horizontal position, others, like these vipers, raise the forepart of the body vertically. At first it was assumed that the dancers were a courting pair, but every time such couples have been

FIGURE 7-2
The combat dance of the Eastern Cottonmouth, *Agkistrodon p. piscivorous*, in a Florida marsh.

collected, both individuals have been found to be males. Whether the dance is an attempt to establish territory or a formal competition for a female has not been determined, but it does seem to have some sexual significance.

Courtship

Many reptiles have a more or less elaborate "Liebespiel" that usually precedes copulation. A male turtle of the genus *Chrysemys* courts a female in the water. Swimming backwards in front of her, or swimming above her, the male vibrates the long claws of his front feet around her cheeks. This is called titillating. Other turtles may bob their heads and nudge and bite at the females.

Among diurnal lizards, particularly those in which the males are brightly marked, the male may posture and display in front of the female. Lizards also indulge in nipping and nudging. Sometimes a pair will walk along, the male straddling the tail of the female and resting his head on her pelvic region when she pauses.

Male and female snakes may glide along side by side, the male caressing the female with his chin and flickering his tongue over her body. He may

FIGURE 7-3
Courtship in the King Cobra, *Ophiophagus hannah,* showing how the male and female glide along side by side. [Courtesy of the New York Zoological Society.]

twine his tail around hers and lie on top of her while his body undulates in a series of waves that pass from his tail toward his head (caudocephalic waves).

Unlike frogs, most reptiles do not have voices and in only a few of them does voice play a part in courtship. Many snakes hiss, and one genus, *Pituophis,* has a special membrane on the epiglottis which vibrates to produce a particularly audible hissing sound when air is expelled from the lungs. Geckos have well-developed voices; some can be heard for distances of nearly a hundred meters. But all of these sounds are apparently warning or threatening, not amatory. The voice of the crocodilian, on the other hand, is definitely associated with breeding, though it is not known whether it is a love call, or a challenge to other males, or both. A fundamental note of 57 vibrations per second typically evokes a roar from a half-grown American Alligator. Indeed, alligators in a lake near the University of Florida are often stimulated to call by fireworks that are exploded on the campus. The primitive New Zealand Tuatara *(Sphenodon punctatus)* has been heard croaking on cold, misty nights, but again the reason for the call is unknown. It is noteworthy that the best-developed reptilian voices are found among nocturnal forms in which hearing is presumably better developed than sight.

Some turtles also have voices, and among the testudines, at least, voice seems to be associated with reproductive activity. The Galapagos tortoise bellows while mating. The male of the South American *Geochelone denticulata* makes a noise while pursuing the female and while copulating that, to anthropomorphic ears, sounds suspiciously like a chuckle.

Copulation

Courtship is usually, but not always, followed by copulation. Sometimes copulation takes place with little or no preliminary "Liebespiel," the male simply grasping the female as soon as he sees her.

Aquatic turtles mate in the water; the others on land. The male approaches the female from behind and mounts her carapace, gripping the front part of her shell with his forefeet, and sometimes biting her head and neck. The hind part of his body is pushed below her carapace. The position is very awkward, particularly for a species with a high, domed shell, and sometimes the male loses his balance and falls.

The positions taken during copulation by crocodilians, lizards, and snakes are much alike. The male lies beside or above the female and thrusts his tail under hers so that the regions of the vents are brought together. Among crocodilians, lizards, and some snakes, the male grasps the head, neck, or shoulder of the female with his jaws. Since a receptive female does not struggle, the grasping of the jaws is probably done simply to help the male keep his balance. Snakes and lizards evert the hemipenis on the side next to the female and insert it in her cloaca. It becomes turgid and, held in place by its spines and calyces, usually cannot be removed forcibly without damage to the animals. Copulation may be completed in a few minutes or it may last for several hours or even a day. Couples in captivity have been known to copulate at repeated intervals for several days.

EGGS AND EMBRYOS

Egg Deposition

The female deposits her eggs several weeks to several months after copulation. The eggs of some reptiles are nearly round, others are ovoid. The shells are either leathery or brittle. The eggs must be laid on land, in places with enough moisture and sufficient heat for incubation. In general, reptiles are more careful in selecting or constructing suitable sites for eggs than most of the amphibians. We see in them a foreshadowing of the elaborate nest-building–incubation–parental-care behavior that characterizes the birds.

The females of some species, such as the Musk Turtle *(Sternothaerus odoratus),* may scatter their eggs in a shallow excavation or leave them only partly hidden in leaf mold or vegetable debris, but most turtles dig fairly deep holes in which to bury their eggs. The female scoops out an excavation with her hind feet, deposits the eggs, then scrapes the dirt back into the hole. She may pound it down with her plastron and carefully scuff over the surface so that all signs of digging are obliterated.

Snakes and lizards usually lay their eggs in or beneath rotten logs, in hollows under rocks, in burrows, or in holes dug in the earth. Sometimes several females will select the same place and thus many eggs will be found together. In Wales, forty bundles of about thirty eggs each were once found in a hole in an old wall. These were eggs of the Ringed Snake *(Natrix natrix helvetica).*

The female American Alligator builds a nest for her eggs. She scrapes together vegetable debris into a large pile as much as two meters in diameter and nearly one meter high. She then scoops a hole in the top, lays her eggs, covers them with some of the material, and packs it. Other crocodilians either build nests or heap piles of sand in which to deposit their eggs.

Some reptiles lay all their eggs at one time, others may lay eggs several times during a breeding season. In cold regions females sometimes reproduce only every other year. In general, reptiles lay fewer eggs than amphibians. The largest clutches, sometimes containing 200 eggs, have been reported for some of the marine turtles. It is noteworthy that these are the forms whose egg-laying area is most restricted. On the other hand, some lizards and snakes produce only one or two eggs at a time. Older females are likely to lay more eggs than younger ones, and larger species more than smaller ones.

Structure of Egg and Embryo

The egg and developing embryo of the American Alligator illustrates both the typical amniote shelled egg, characteristic of the reptiles, and some of the ways the reptilian egg differs from the more familiar hen's egg.

Alligator eggs vary in size, ranging from 65 to 85 mm in their longest diameter and from 38 to 50 mm in their shortest. At first they are slimy and pure white, but they generally become stained by the damp, decaying vegetation of which the nest is built. The shell is thicker and coarser than that of the hen's egg. It is very calcareous and is readily dissolved in dilute acids. Inside the shell are two indistinct layers of shell membranes, but no air chamber like that in a hen's egg forms between them at any time during development.

Shortly after it is laid, a typical alligator egg develops a white, chalky band around the lesser circumference; the band usually runs entirely around the egg, and varies in width from 15 to 35 mm. The band is caused by the development of a chalky substance in the shell membranes. This band is thought to aid in the passage of gases to and from the developing embryo.

The white of an alligator's egg is much denser than the albumen of a hen's egg. The entire egg can be emptied from the shell and passed from one hand to the other without serious danger of rupturing either the mass of albumen or the enclosed yolk. The albumen is apparently not divided into layers of decreasing density as it is in the hen's egg. Its color varies from a pale yellowish-white (the most usual color) to a definite green. The albumen of the crocodile is normally light green. The yolk is a pale yellow, spherical mass lying in the center of the white. Its diameter is so great that it comes very close to the shell around the small circumference of the egg, where it is covered only by a thin layer of white. Care must be taken not to rupture the yolk when removing the shell from this portion. The yolk is very fluid and is contained in a delicate vitelline membrane.

Development starts in the oviduct and by the time the egg is laid a definite young embryo has formed. The amnion develops rapidly and entirely from the anterior end of the embryo. The relationship between embryo, yolk mass, amnion, allantois, and chorion is illustrated in Figure 3-1. Like the embryo of the chicken, the alligator embryo shows torsion or twisting of the body, so that the anterior end lies on its side, but unlike the embryo of the chicken, which always twists to the right, the embryo of the alligator may turn either to the left or to the right.

Just before hatching, the young alligator is about twenty cm long, nearly three times the length of the egg, but the compressed tail is bent so that, although it forms about half the length of the animal, it really takes up little room within the egg. The umbilical cord is withdrawn, but the scar is still present at the time of hatching.

Parental Care

Most reptiles abandon their eggs as soon as they are laid. Indeed, the mother may never even see them. Reports of a female guarding her clutch, based simply on the chance observation of an individual in the vicinity of the eggs, should be viewed with skepticism. Nevertheless, there is evidence that at least a few reptiles have developed care-giving behavior beyond that shown in the preparation of a proper incubation site.

Female skinks of the genus *Eumeces* stay with their eggs. The mother protects them from small predators, and may turn the eggs periodically and bring them back together if the clutch is accidentally scattered. It has even been suggested that she helps to incubate them by basking in the sun and then coiling her warmed body around them.

Females of several species of snakes are known to coil about the incubating eggs. Perhaps best known are the pythons. After depositing the eggs, the female python draws them together by moving her tail and body. She heaps them in a pyramidal mass and coils around them, her head capping the spiral formed by her body. She stays with the eggs for six weeks, only rarely moving away to drink, but leaves them about two weeks before they hatch. A brooding female python is able to raise her body temperature as much as $7.3\,^{\circ}C$ above the ambient temperature by spasmodic muscular contractions. In her ability to carry on thermoregulation by endogenous heat production, the brooding python thus resembles the endothermic birds and mammals.

Other snakes that are known to guard their eggs include the cobras and the American Mud Snake *(Farancia abacura)*.

The best-developed care-giving behavior pattern so far reported for a reptile is that of the American Alligator. During the incubation period of nine or ten weeks the mother remains in the vicinity of the nest, and is said to moisten it occasionally with the contents of her bladder. When the young hatch they start a high-pitched grunting that attracts the attention of the mother. If the surface of the nest is packed too hard, she may scrape it away to release the young. She is said to scoop out a wallow pool for them, and to guard them from attack by predators, including other alligators. The young may remain with the mother for a year or more.

One of the most prevalent and persistent of all snake myths is the story of the mother swallowing her young to protect them and later disgorging them. The tale, which is usually told of ovoviviparous species, is of course untrue, but it may have some basis in fact. In captivity the young of some of these species (for example, the European Viper) have been seen to remain in the vicinity of the mother for a couple of days after birth and disappear beneath her body when threatened. Whether they do the same in the wild is not certain, but if they do, such an association must be at most an ephemeral one.

HATCHING, BIRTH, AND YOUNG

Heat is necessary for the development of vertebrate embryos, and the rate of development is in part determined by the amount of heat available. The endothermic birds and mammals provide a relatively high, constant source of heat for their developing young, either by retaining them in the body of the mother, or by setting on the eggs. The length of the embryonic period is fairly constant for each species, though it varies, of course, from one species to another. In contrast, eggs of most ectothermic reptiles, like their parents, are dependent on environmental heat, and since this may vary widely from year to year and from place to place, the length of time it takes reptile eggs to hatch, even within a single species, is extremely variable. The hotter the summer, the sooner the young reptiles appear. In France, young of the Smooth Snake *(Coronella austriaca)* usually hatch in late August or early September, but if the summer is cold they may not appear until October. In the northern United States and Canada, young Painted Turtles *(Chrysemys picta)* may winter in the eggs and hatch the following spring.

Even in live-bearing reptiles, the gestation period is variable. Pregnant females spend much time basking in the sun and the young are born earlier in a warm summer. In England, young of the Slow-worm (a lizard, *Anguis fragilis*) and of one species of Viper *(Vipera berus)* may remain in the body of the mother throughout the winter that follows a cool summer.

Eggs of the Tuatara *(Sphenodon punctatus)* are said to take thirteen months to develop. Table 7-3 gives some indication of the range of incubation periods that have been reported for several reptiles.

TABLE 7-3

Length of Incubation Period
for Various Species of Reptiles

Species	Incubation period (days)
Caretta c. caretta	31–65
Chelodina longicollis	90–120
Chelydra serpentina	81–90
Eumeces fasciatus	28–49
Lacerta a. agilis	49–84
Sceloporus undulatus	70–84
Alsophis angulifer	89–97
Coluber constrictor mormon	51–64
Elaphe guttata emoryi	72–77
Lampropeltis getulus californiae	66–83
Naja naja	69–76
Pituophis m. melanoleucus	59–101
Pituophis m. annectens	64–77

A young reptile that is ready to hatch needs some means of cutting its way from the egg. Embryonic turtles and crocodilians have a horny projection called the "caruncle" on the tip of the snout, with which they pierce or slash the shell. Lizards and snakes bear an egg tooth on the front of the premaxillary bone in the upper jaw. An egg tooth is shed within a day or two after birth, but a caruncle may persist for weeks. In the young of ovoviviparous species, which are usually born encased in a thin membrane that can easily be ruptured by the snout, and is sometimes broken before the young leave the body of the mother, the egg tooth has degenerated.

FIGURE 7-4
Hatchling Mountain Patch-nosed Snakes, *Salvadora grahamiae*, as they emerge from the eggs.

Young reptiles resemble their parents in both structure and behavior. Baby vipers and crotalids have well-developed, functional fangs; baby cobras can spread their hoods. The skin is first shed shortly after birth. Similar as they are to the adult reptiles in morphology, the young often differ strikingly from their parents in color. They are frequently much more brightly marked and sometimes have entirely different color patterns. Adult Black Racers *(Coluber c. constrictor)* are a smooth, velvety black above, but the young are vividly marked with brown saddles and spots on a tan background.

GROWTH, MATURITY, AND LONGEVITY

The young grow rapidly at first, then more slowly. Some reptiles apparently reach a definite size and stop growing; others grow, although very slowly, throughout their lives. As human children do, reptiles seem to grow in spurts, with alternating periods of rapid and slow growth. During hibernation growth all but stops.

Reptiles are ready to breed before they are full grown. For some, at least, sexual maturity seems to be more a matter of reaching a certain size than a certain age. Males of the Red-eared Turtle *(Chrysemys scripta elegans)* are sexually mature when the plastron is 9 or 10 cm long, but it may take individuals from two to five years to reach this size. Females mature at a plastral length of 15 to 19.5 cm, which is attained in three to eight years. American Alligators do not breed until they are nearly 2 meters long and

TABLE 7-4
Longevity of Some Species of Reptiles

Order, family, and species	Maximum age (years)
Testudinata	
Chelydridae *(Macrochelys temmincki)*	59
Testudinidae *(Testudo sumeirei)*	152
Cheloniidae *(Caretta caretta)*	33
Trionychidae *(Trionyx triunguis)*	25
Pelomedusidae *(Pelusios derbianus)*	41+
Chelidae *(Chelodina longicollis)*	37+
Rhynchocephalia	
Sphenodontidae *(Sphenodon punctatus)*	77
Squamata	
Lacertilia	
Gekkonidae *(Tarentola mauritanica)*	7
Iguanidae *(Conolophus subscristatus)*	15
Agamidae *(Physignathus lesueuri)*	6
Scincidae *(Egernia cunninghami)*	20
Cordylidae *(Cordylus giganteus)*	5
Teiidae *(Tupinambis teguixin)*	13
Anguinidae *(Anguis fragilis)*	54
Helodermatidae *(Heloderma suspectum)*	20
Varanidae *(Varanus varius)*	7
Serpentes	
Boidae *(Eunectes murinus)*	29
(Constrictor constrictor)	25
Colubridae *(Elaphe situla)*	23
(Drymarchon corais couperi)	25
Elapidae *(Naja melanoleuca)*	29
Viperidae *(Vipera ammodytes)*	22
(Agkistrodon piscivorus)	21
(Crotalus atrox)	22
Crocodilia	
Crocodilidae *(Alligator mississippiensis)*	56

six or seven years old. Many lizards and snakes take about three years to reach sexual maturity. There is some indication that tropical forms, whose growth is not interrupted by periods of hibernation, mature more quickly. Among the Javanese snakes, the Blunthead *(Pareas carinatus)* is usually ready to breed in 11 months, the Red-necked Keelback *(Rhabdophis subminiata)* in 13 months, and the Greater Rat Snake *(Ptyas mucosa)* in 20 months.

As with amphibians, our knowledge of the maximum age to which reptiles live is based largely on specimens kept in captivity. Table 7-4 shows the longevity records for a number of species. Apparently, turtles tend to live longer than snakes, and snakes live longer than lizards. Again there seems to be a negative correlation between the degree of activity and the duration of life.

Our knowledge of the reproductive habits of many species of reptiles is either a complete blank or consists solely of a record of a single clutch of eggs discovered by chance. Yet it is clear that, in spite of the variations in detail, the over-all pattern of reptilian reproduction is much less diversified than that of the amphibians: fertilization is always internal; if eggs are laid, they are always laid on land; development is always direct.

The diversity shown by the modern reptiles is just that which must also have been present in the ancestral forms to have allowed the evolution of the two great groups that arose from them. The birds specialized in the development of the hard-shelled egg, well supplied with albumen; the higher mammals specialized in viviparity.

READINGS AND REFERENCES

Gadow, H. *Amphibia and Reptiles.* Cambridge Natural History, vol. 8. London: Macmillan, 1901.

McIlhenny, E. A. *The Alligator's Life History.* Boston: Christopher, 1935. (Based on first-hand experience with our most spectacular native reptile.)

Mertens, R. *The World of Amphibians and Reptiles.* Translated by H. W. Parker. New York: McGraw-Hill, 1960.

Oliver, J. A. *North American Amphibians and Reptiles.* Princeton, N.J.: Van Nostrand, 1955.

Parker, H. W. *Snakes.* New York: W. W. Norton, 1963.

———. *Natural History of Snakes.* London: British Museum (Natural History), 1965.

Reese, A. M. *The Alligator and Its Allies.* New York: Putnam's, 1915.

Schmidt, K. P., and R. F. Inger. *Living Reptiles of the World.* Garden City, N.Y.: Hanover House, 1957. (A beautifully and profusely illustrated account of the families of reptiles, including much life history data.)

Smith, M. A. *Fauna of British India. Reptilia and Amphibia.* Vols. 1–3. London: Taylor and Francis, 1931-1943. (Contains much life history information.)

———. *The British Amphibians and Reptiles.* Rev. ed. London: Collins, 1954.

8

HOMEOSTASIS

Like other animals, the amphibians and reptiles must maintain a certain degree of constancy of their internal environment. Thus, the extra-cellular fluids must be kept in osmotic equilibrium with the cells; pH levels must be maintained within narrow limits; the body temperature, at least during periods of activity, must be neither too high nor too low for the proper functioning of the various enzyme systems. The maintenance of the constancy of the internal environment is known as "homeostasis." Homeostasis involves both internal control mechanisms that are largely regulated by hormones and the ability of the animal to exchange energy and material with its environment.

The study of the internal control of homeostasis falls properly into the field of experimental physiology. The exchange with the environment, on the other hand, since it involves the whole animal in its habitat, falls properly within the realm of herpetology.

In this chapter we shall be concerned with three of the major commodities of exchange: heat, respiratory gases, and water. As might be expected, the reptiles, with their dry, scaly integument, have problems in gaining or losing water and chemicals in solution in water (i.e., gas, nitrogenous waste, salts, etc.) that differ from those of the amphibians.

HEAT

Heat is produced in the animal body as a by-product of metabolic activity. The respiratory and circulatory systems of the herptiles are less efficient than those of the birds and mammals in bringing an abundant supply of oxygen to the tissues for metabolic activity. Since their metabolic rate is consequently low, the herptiles do not produce enough internal heat to maintain body temperatures high enough for an optimal existence. They must absorb additional heat from an outside source: the sun's rays, or the surrounding medium, or the surface on which they rest. Nocturnal herptiles draw upon the solar heat absorbed by the substratum during the day. In short, the herptiles are ectothermic.

Mechanisms of Temperature Control

To say that the herptiles absorb heat from the environment does not mean that their body temperatures are the same as those of the environment. The Squamata, at least, show a remarkable ability to control the level of body temperature during their periods of activity and they may be considerably warmer than the surrounding air. An iguanid lizard, *Liolaemus m. multiformis*, found at a high altitude in southern Peru, had a cloacal temperature of 31°C, although the air temperature in the shade nearby was only 0°C. A lizard may bask in the sun until it is sufficiently warmed, then, to prevent its body temperature from going too high, may move into a burrow or den, into water, or simply into a shady spot. A Sidewinder *(Crotalus cerastes)* was observed to maintain its temperature between 31° and 32°C by simply exposing more or less of its body to direct solar radiation while coiled at the mouth of its burrow.

The ability of some herptiles to change color may serve as a thermoregulatory mechanism, since a dark color absorbs heat rays, and a light one reflects them. Lizards of the genus *Anolis* in Louisiana are green in the high temperatures of a summer day (27° to 32°C) but in the cooler parts of the day in other seasons most are brown.

Panting is essentially a mechanism that conflicts with lung inflation and under normal circumstances occurs during times when breathing is suspended (apnea). It functions primarily to disperse body heat and is known to be effective in lowering body temperature or in slowing its rate of increase in the lizards *Dipsosaurus, Crotaphytus*, and *Iguana*.

In addition to behavioral control of heat exchange with the environment, some reptiles exhibit a physiological control of their body heat. Among lizards this has been shown in the families Scincidae, Agamidae, Varanidae and Ignanidae. This control involves both endogenous heat production, which accelerates body heating and retards body cooling, and cardiovascular adjustments, which can modify heat transport within the body.

Perhaps the best known example of endogenous heat production in reptiles is that of the Indian Python, *Python molurus bivittatus.* A brooding female can maintain a body temperature as much as 7.3°C higher than the ambient air temperature by palpable spasmotic body contractions. Individuals of the same species in nonbrooding condition maintain a body temperature only 1° or 2°C above the ambient air and substrate temperature.

Critical and Eccritic Temperatures

The critical minimum temperature is the point at which an animal suffers from cold torpor to the degree that it is helpless to escape from enemies or to act to raise its body temperature. The critical maximum temperature is that at which locomotor activity becomes disorganized and the animal becomes incapacitated. These are ecological lethal points, since, although the temperatures alone may not kill, they prevent the animal from performing its necessary life-preserving activities.

Between the critical minimum and the critical maximum, a reptile has an eccritic temperature to which it tends to adjust its body temperature during periods of activity by thermoregulation. The Desert Spiny Lizard *(Sceloporus m. magister)* feeds well and seems superficially to thrive at a body temperature of approximately 30°C, but defecates frequently only at body temperatures of about 37° to 38°C; this higher temperature may be necessary to bring about peristalsis; therefore the lizard probably thrives best at the higher temperature. These eccritic or "preferred" temperatures of active reptiles are surprisingly close the the critical maxima, which in most reptiles are only about 5° to 6°C higher.

It should be noted, however, that the temperature to which a lizard will adjust in a thermal gradient in the laboratory is not necessarily the same as the one it will seek in nature. In the field the animal may adjust to a temperature higher or lower than it does in the laboratory. Neither does a lizard invariably adjust to a temperature at which it seems to carry on metabolic activities to the optimum extent. Specimens of the Island Night Lizard, *Klauberina,* in a gradient chamber over a 24-hour period voluntarily selected such low temperatures during the night that they became torpid, whereas they selected higher temperatures during the daylight hours.

Studies of a group of diurnal lizards *(Sceloporus occidentalis, S. undulatus, Uta stansburiana,* and *Uma notata)* indicate that the parietal eye (a rudimentary third eye found in the center of the forehead in *Sphenodon* and some lizards) helps regulate the amount of exposure to sunlight. Lizards whose parietal eye was removed or shielded by aluminum foil spent more time

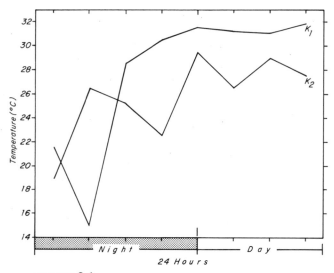

FIGURE 8-1

Temperatures voluntarily selected by two individuals of
Klauberina riversiana kept in a gradient chamber for a
period of twenty-four hours. [Redrawn from Regal,
Science, 1967.]

exposed to light and remained on the surface of the ground longer than did
untreated animals. Consequently, although the eccritic temperatures were
the same for both groups, the treated lizards maintained the thermal levels
necessary for normal activity for longer periods during the day. They were
also less viable than the untreated ones when deprived of food; this was
probably because they used up their available energy faster.

In the California desert, diurnal lizards maintain daytime body temperatures of about 37°C. Nocturnal lizards not only tolerate but actually prefer
lower body temperatures. Their eccritic temperatures are roughly 29° to
30°C. Body temperatures of snakes are usually a few degrees lower than
those of lizards, and again the nocturnal types have lower eccritic temperatures than do the diurnal ones. For example, the Racer *(Masticophis flagellum
piceus)*, a diurnal snake, has an average activity temperature of 33°C,
whereas the nocturnal Glossy Snake *(Arizona elegans occidentalis)* of the
same region, has a normal activity temperature of about 27°C.

Species within a genus have about the same normal temperature range
even though they may be quite widely separated geographically and cli-
matically, whereas the temperatures of two different genera may be quite
dissimilar even in almost identical habitats. The Desert Spiny Lizard *(Sceloporus m. magister)* of the American Southwest has body temperatures ranging
from 31° to 38°C, but those of the Checkered Whiptail *(Cnemidophorus*

tessellatus) from the same region range between 37° and 44°C. On the other hand, the Florida Scrub Lizard *(Sceloporus woodi)* has body temperatures ranging from 31° to 39°C, and the Six-lined Racerunner *(Cnemidophorus sexlineatus)* from Florida ranges from 38° to 44°C. Thus the two species of *Sceloporus* have closely similar body temperatures and so do the two *Cnemidophorus*. It is evident that these lizards have the ability to regulate their thermal levels enough to keep them near their eccritic temperatures regardless of the habitat in which they are found.

The "coldest" reptile is the sluggish, nocturnal Tuatara *(Sphenodon),* which lives on cold, windy, foggy islands off the coast of New Zealand. Body temperatures of specimens taken in the wild at night have ranged between 6° and 13.3°C. *Sphenodon* has an extremely low metabolic rate, and although it can move fairly rapidly for short distances, it is probably incapable of sustained activity comparable to that of the lively lizards.

Male sex cells seem to be more sensitive to the effects of heat than other body tissues. Geographic variation in spermatogenic activity of the Garter Snakes *(Thamnophis)* can be correlated with seasonal increases and decreases

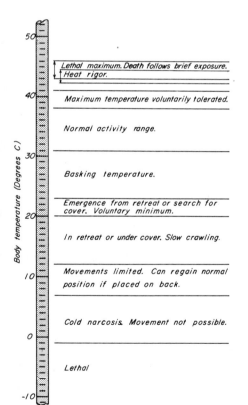

FIGURE 8-2
The significance of body temperature to the behavior of reptiles is indicated by this chart, which gives the approximate temperatures for various activities of Spiny Lizards *(Sceloporus)*. The effects of exposure to heat levels near the extremes depend on the duration of exposure. Even temperatures near the upper limit for the lizards' normal activity become lethal if exposure is prolonged. [From "How Reptiles Regulate Their Body Temperature" by Charles M. Bogert. Copyright © 1959 by Scientific American, Inc. All rights reserved.]

134

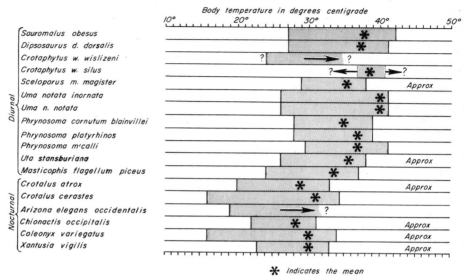

FIGURE 8-3

Normal activity ranges of 18 reptiles of the southwestern United States, including both snakes and lizards. Twelve diurnal forms are arranged roughly in the order of size, with the larger species at the top of the column, and six nocturnal forms are arranged in the same fashion. Stippled portions of the bars indicate the normal activity range, that is, the temperatures between the minimum and maximum voluntarily tolerated by individual species under conditions (1) in the field, (2) in cages set up in the field, or (3) under laboratory conditions. The asterisk (*) indicates the mean for temperatures recorded when individuals were engaged in normal activities, and hence represents an approximation of the ecological optimum. Data for all species are not sufficiently extensive to be considered conclusive. Consequently portions of this chart are provisional, and revisions may be necessary after controlled experiments in the laboratory have been conducted for the forms in question.
[After Cowles and Bogert, 1944.]

in the number of hours a day the snakes can maintain their body temperatures within the normal activity range. It has been suggested that increase in temperature causes sterility in the male Desert Night Lizard *(Xantusia vigilis)*. Strangely enough, such correlations have not been found in the females.

Like the reptiles, amphibians apparently have preferred body temperatures. They are able to exercise some behavioral thermoregulation, though this control is less precise than that of the reptiles. Salamanders seek local situations within the habitat in which the temperature is appropriate. Most salamanders that have been tested have body temperatures below 20°C. Active specimens of the Slimy Salamander *(Plethodon glutinosus)* of the eastern United States have body temperatures ranging from 16° to

20.3 °C; the lowest on record, 7.1 °C, is that of the Mount Lyell Salamander *(Hydromantes),* which lives at the borders of the snow line.

The body temperature of a frog is generally higher than that of a salamander. On cloudy days the body temperature of a frog is usually close to that of the environment, but in clear weather individuals of some species will bask in the sun and may raise their temperatures as much as 10°C above that of the surrounding air or water.

Temperature and Distribution

Temperature undoubtedly plays an important role in determining the geographic distribution of the herptiles. They are limited to regions where sufficient environmental heat is available at least part of the year. The persistence of so many primitive species on islands and peninsulas may result in part from the equable maritime climate of these land forms. Islands do not present as rigorous an environment, so far as temperature fluctuations are concerned, as do the continental masses. The "temperate" climates of the large land masses, particularly outside of the tropics, are really most intemperate. In the later geologic epochs, these areas have been characterized by climatic extremes to which the primitive forms may have been unable to adjust. As a result, these species have been eliminated on the continents but have persisted on islands.

GAS EXCHANGE

Probably no group of vertebrates exemplifies the evolution of gas transport systems better than do the herptiles.

Oxygen is more readily available to air-breathing animals than to water-breathing ones; for example, the oxygen environment provided by water at 20°C is twelve times more rarefied than the atmosphere at the top of Mount Everest. Water-breathing amphibians such as the trachystomes and salamanders like *Necturus* have to pump a great deal of water over the gills to obtain the necessary O_2. Terrestrial animals need to pump much less air into their lungs to obtain an equivalent amount. On the other hand, because carbon dioxide is much more soluble in water than O_2 is, water at 16°C is as effective as air in the dispersal of CO_2 from the respiratory surfaces. Water flowing over the gills at a rate that will provide the O_2 needed by an aquatic animal provides also for the rapid removal of CO_2 from the system. The air flow over the lung surfaces that provides the requisite O_2 removes much less of the CO_2. As a result, the problems of gaining O_2 and dispersing CO_2 are quite different in aquatic, semiaquatic, and terrestrial animals.

Respiratory movements may be stimulated either by oxygen depletion or by carbon dioxide stress. The respiratory center in the brain is sensitive to CO_2 and initiates respiratory movement if the CO_2 concentration in the body fluids exceeds a certain level. In frogs, the carotid "gland," a spongy mass located at the base of the internal carotid artery, is sensitive to oxygen concentration in the blood and, if the O_2 is lowered, will stimulate respiratory movements. It has been shown in *Rana* that CO_2 stimulates respiratory movements even after the carotid gland has been obliterated but that O_2 deficiency acts as a stimulant only if the carotid gland is intact.

Gas Exchange in Amphibians

Amphibians may exchange gas with the environment by means of gills, lungs, buccopharyngeal mucosa, or the skin. For air-breathing, lunged salamanders (like *Ambystoma and Taricha*) oxygen absorption by the lungs increases linearly with an increase in temperature whereas absorption through the skin remains essentially constant at all temperatures. At 15°C

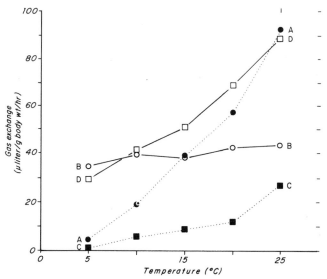

FIGURE 8-4
Gas exchange in the lunged salamander *Taricha granulosa*.
Gas exchange in microliters per gram of body weight per hour
is plotted against temperature. Line *A–A* represents pulmonary
O_2 (in lungs, or buccopharyngeal mucosa, or both; *B–B*,
cutaneous O_2; *C–C*, pulmonary CO_2; *D–D*, cutaneous CO_2.
[Redrawn from Whitford and Hutchison, *Physiol. Zool.*, 1965.]

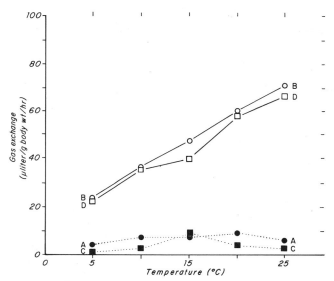

FIGURE 8-5

Gas exchange in the lungless salamander *Desmognathus quadramaculatus.* For explanation, see legend of Figure 8-4. [Redrawn from Whitford and Hutchison, *Physiol. Zool.*, 1965.]

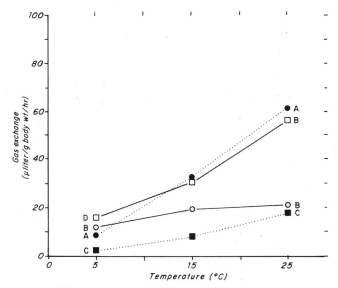

FIGURE 8-6

Gas exchange in the Green Frog, *Rana clamitans.* For explanation, see legend of Figure 8-4. [Redrawn from Vinegar and Hutchison, *Zoologica*, 1965.]

the skin accounts for approximately 50 percent of the O_2 intake, but at 25°C more than twice the amount of oxygen is taken in through the lungs as through the skin.

Among the lungless salamanders, on the other hand, the skin is more important for O_2 transport. At no temperature does the buccopharyngeal mucosa account for as much O_2 intake as the skin in *Desmognathus quadramaculatus* and at 15°C 85 percent of the oxygen is taken in through the skin. In the more terrestrial plethodontid *Plethodon glutinosus,* the amount of oxygen taken in through the skin at 15°C is 76 percent. Pulmonary and cutaneous O_2 uptake are about equal in *Rana clamitans* at 8°C but at 25°C pulmonary uptake of O_2 is about three times that of the skin.

Respiratory movements in the frog include cycles of buccal oscillations and lung ventilation. During the buccal oscillations the nostrils are open, the glottis is closed, and the floor of the mouth is alternately raised and lowered; this brings fresh air into the buccal cavity. At the start of lung ventilation, the floor of the mouth is lowered and the glottis is opened. Since the air in the lungs is above atmospheric pressure, a stream of air flows from the lungs through the mouth and out the nostrils. The greater part of the buccal cavity lies below the opening of the glottis and there is little mixing of the air in this part of the chamber with the air being expired. After expiration the nostrils are closed, the floor of the mouth is raised, and the buccal air is pumped into the lungs.

Some, but not all, air-breathing amphibians continue buccal movements while submerged. Apparently some O_2 is absorbed from the water in such cases. The time that such an amphibian can survive under water is probably dependent upon how well the buccopharyngeal membranes are adapted to picking up O_2. In water at 15°C *Rana pipiens* can carry on buccal respiratory movements for as long as seven hours while the more terrestrial *Bufo terrestris* expires after three hours.

Since the relative proportions of the gases in the air remain constant at different heights, one of the most important results of the decreased air pressure at high altitudes is the decrease in the total amount of oxygen available. Oxygen is transported by hemoglobin in red blood cells from the lungs to the tissues; hemoglobin is measured in parts per hundred by weight of the blood. To say, for example, that a frog has a hemoglobin value of 10 means that the weight of the hemoglobin makes up 10 percent of the total weight of the blood. Frogs from the lowlands have lower hemoglobin values than those from mountains. Hemoglobin values ranging from 7.5 to 13.5 have been reported for frogs and toads from the European lowlands, but specimens of the toad *Bombina variegata,* from the Alps, have values ranging from 14.25 to 15.25. In Guatemala, *Bufo marinus,* which occurs from sea level to elevations of about 1500 meters, has a mean hemoglobin value of 8.66, whereas *Bufo bocourti,* which ranges from about 1700 to 3400 meters has a mean value of 10.77. If the hemoglobins of these species have similar oxygen-carrying capacities (which is not yet known), the increase in hemoglobin could balance the decrease in oxygen at the higher altitudes.

Although the primitive amphibian lung is fairly adequate for the purpose of maintaining oxygen tension it is inadequate for keeping the CO_2 tension at a low level. Instead the skin is utilized to cope with the rising level of CO_2 brought about by breathing air. *Taricha granulosa* disperses 80 percent of the CO_2 through the skin at 15°C, and the percentage is only slightly lower at higher temperatures. In the lungless *Desmognathus quadramaculatus* the skin accounts for 75 percent of CO_2 disposal at 15°C and as much as 95 percent at 25°C. *Rana clamitans* loses about 80 percent of the CO_2 through the skin at all temperatures.

Gas Exchange in Reptiles

The reptiles were enabled to colonize land not only by the development of the amniote egg but also by the development of a relatively water-impervious integument. Respiratory gas exchange through the skin was no longer possible, and even though the improved reptilian lung allowed sufficient O_2 uptake it did not allow for sufficient elimination of CO_2. Physiological adaptations were thus required to make it possible for the animals to tolerate higher CO_2 concentrations in their body fluid than were present in the amphibians with their permeable skins through which most of the CO_2 was dispersed. Of the CO_2 that enters the bloodstream most combines with water molecules to form carbonic acid (H_2CO_3) which then dissociates to form bicarbonate ions (HCO_3^-). Adaptation of the reptiles to a fully terrestrial existence required adjustment to higher levels of bicarbonate ions in the blood.

In the truly terrestrial reptiles three types of respiratory movement occur: lung ventilation, buccopharyngeal pumping, and panting. Air is pumped into the amphibian lung by buccal force, but is drawn into the reptile lung by suction. Most reptiles take in air by a muscular expansion of the rib cage and body wall. Since this lowers the pressure in the lungs and abdominal cavity below atmospheric pressure, air moves into the lungs. There is then a respiratory pause with the glottis closed and the lungs inflated. Air is forced from the lungs by the elastic recoil of the lung tissue and by contractions of the body-wall musculature. The turtle, with its rigid shell to which the ribs are fused, changes the size of the body cavity, and hence the intracavity pressure, by alternate contractions of antagonistic muscles in the flank and shoulder region.

It was thought for a long time that buccopharyngeal pumping in most reptiles functioned to ventilate the lungs, but it is now known that at least in the lizards and turtles pharyngeal pumping takes place only during respiratory pauses and functions primarily as an olfactory mechanism.

Many of the aquatic turtles are known to be able to stay submerged for long periods without breathing. In turtles of the genera *Trionyx* and *Sternothaerus* enough oxygen is gained from pumping water in and out of the pharynx to maintain metabolic processes for long periods if the turtle is not physically active. Thus the animal may be able to stay submerged indefinitely. Not all diving turtles have this ability though, as *Chrysemys scripta* and related species seem to be incapable of deriving appreciable O_2 from the water.

Among reptiles, turtles seem especially adapted to tolerate anoxia. Table 8-1 shows the mean time in minutes that members of different families of reptiles are known to survive in an atmosphere of nitrogen.

TABLE 8-1

Tolerance of Anoxia in Various Families of Reptiles

Order and family	Number of species tested	Mean time (minutes)
Testudinata		
Chelydridae	1	1050
Testudinidae	14	945
Pelomedusidae	2	980
Kinosternidae	5	876
Trionychidae	1	546
Chelidae	2	465
Cheloniidae	2	120
Squamata		
Lacertilia		
Iguanidae	6	57
Gekkonidae	1	31
Anguinidae	1	29
Scincidae	4	25
Teiidae	1	22
Serpentes		
Viperidae	3	95
Boidae	3	59
Elapidae	1	33
Colubridae	22	42
Crocodilia		
Crocodylidae	1	33

Source: Data from Belkin, *Science*, 1963.

The ability of aquatic turtles to stay submerged involves physiological reactions in addition to the ability to obtain O_2 from the water. These reactions include inibition of heart beat, anaerobic glycolysis (to provide energy), increase in plasma bicarbonate, and possibly the use of hydrogen acceptors other than pyruvic acid in glycolysis. The turtles other than the sea turtles have approximately two and a half times as much plasma bicarbonate as other reptiles: sea turtles have half again as much.

During typical anaerobic respiration in animals, the pyruvic acid that results from the breakdown of glucose accepts hydrogen from the hydrogen acceptor, NAD, and is thereby converted into lactic acid. The possibility that turtles are able to use hydrogen acceptors other than pyruvic acid in glycolysis should not be discounted although it remains to be tested. Carp, when exposed to long-term anoxia, form fat rather than lactic acid and it seems probable that turtles may also be able to form substances less disturbing to the tissues than lactic acid.

WATER BALANCE

Because animals that live in fresh water have body fluids that are hyper-osmotic to the surrounding medium, water tends to move into their bodies by osmosis. On the other hand, animals that live in the sea usually are hypo-osmotic to sea water and thus tend to lose water to the environment. Terrestrial forms also lose water to the surrounding air. In addition to the water needed to keep the body in osmotic balance, there must also be water available to excrete nitrogenous wastes. These wastes are predominantly ammonia, urea, and uric acid. Ammonia is highly water soluble and very toxic, urea is also highly water soluble but less toxic than ammonia, and uric acid is relatively insoluble and nontoxic.

Osmoregulation in Amphibia

The amphibians, with their diverse habitats and life histories, have varying problems of water balance and have met these problems in diverse ways. In general, osmoregulation is achieved by a balance between water uptake through the skin and water loss through the kidneys.

Most amphibian larvae develop in fresh water. They produce copious amounts of dilute urine to eliminate the excess water that moves into their bodies by osmosis. Their nitrogenous wastes are mostly in the form of ammonia, which is rapidly flushed from the body in the urine. During metamorphosis from the aquatic larva to the terrestrial adult there is a sudden, dramatic shift in the form in which nitrogen is excreted. Urea becomes the chief excretory product. The bullfrog, *Rana catesbeiana*, excretes 84 percent of its nitrogenous wastes in the form of urea as an adult, but only 10 percent as a tadpole. Amphibians that remain aquatic as adults usually retain the larval excretory pattern. The aquatic African Clawed Frog, *Xenopus laevis*, excretes about 75 percent ammonia and 25 percent urea as a tadpole.

During metamorphosis the excretion is 46 percent ammonia and 54 percent urea, but the adult reverts to the tadpole pattern. Since urea is less toxic than ammonia, it does not need to be removed so rapidly from the body, and the terrestrial amphibians are able to reduce their rate of urine production during periods of water shortage.

The amphibian skin is not universally perfectly water permeable as is often assumed. Many frogs, particularly the more terrestrial species, have a definitive layer of ground substance (layer G) between the stratum compactum and stratum spongiosum of the dermis. This noncellular layer, which has been shown by straining to consist of calcium and acid mucopolysaccharides (polysaccharides with attached amino groups), presumably plays an important part in defense against desiccation.

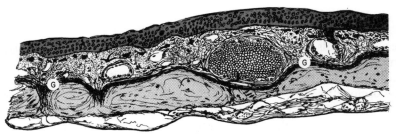

FIGURE 8-7
Layer G as it might appear in *Rana arvalis*. [Redrawn from Elkan, 1968.]

Although the exact method by which layer G functions to control water transport is not known, that it does so is implied by its development in different species of frogs. Layer G is essentially absent in Ascaphidae, Pipidae, and the aquatic leptodactylids, but it is well developed in most Hylidae, the more terrestrial leptodactylids, and in all adult Pelobatidae, Bufonidae, and Ranidae that have been examined—a total of 42 species in the last three families. All salamanders examined so far seem to lack layer G.

Temperature affects water balance in frogs. Skin permeability varies directly with temperature. Urine flow decreases 2 or 3 fold for each 10°C drop in temperature between 24° and 0°C. *Rana clamitans* immersed in water in a cool room (2° to 5°C) produced no urine for a period of 12 to 18 hours although body weight increased about 6 percent in the same period as a result of the uptake of water through the skin. When urine production resumed it remained at about 25 percent of the level at which it forms at 20° to 25°C.

Under stress of desiccation glomerular filtration diminishes. In *Rana clamitans* placed in relatively dry laboratory air at room temperature the filtration rate diminished by just a few percent during the first hour, but it fell to half the control level during the second hour.

Amphibians vary considerably in their ability to tolerate dehydration; this seems to be definitely associated with habitat preference. Ten species of frogs were tested in a dehydration chamber in which artificially dried air was passed over them. The frogs were weighed before and after dehydration; the weight change was equal to the amount of moisture lost during dehydration (see Table 8-2).

TABLE 8-2

The Vital Limits of Water Loss for Ten Species
of Frogs from Various Types of Habitats

Species	Habitat	Loss in body weight (%)
Scaphiopus hammondi	Terrestro-fossorial	48.8
Scaphiopus holbrooki	Terrestro-fossorial	48.1
Bufo boreas	Terrestrial	43.6
Bufo terrestris	Terrestrial	43.0
Hyla regilla	Terrestrial	39.0
Hyla cinerea	Terrestro-arboreal	37.3
Rana pipiens	Terrestro-semiaquatic	36.6
Rana aurora	Semiaquatic	34.0
Rana sphenocephala	Semiaquatic	32.4
Rana grylio	Aquatic	31.2

Source: Thorson and Svihla, Ecology, 1943.

The table shows that frogs occupying dry habitats tolerate dehydration better than those occupying more humid habitats. Spadefoots (Scaphiopus) are notably animals of dry land and can stand a loss of body weight of almost 50 percent. The most aquatic species on the list is the Pig Frog (Rana grylio), characteristically found sitting on lily pads or floating at the surface of the water in marshes of the southeastern United States. It could tolerate a loss of body weight of less than one-third.

Amphibians have various ways of coping with a reduced amount of environmental moisture. A Climbing Salamander, Aneides, coiled itself in a dehydration chamber, and thereby reduced the amount of body surface it exposed to the dry air.

Although it has been stated in the literature that water gain in frogs is due entirely to intake through the skin, this is not always so: two species of Andean Gastrotheca have been observed to drink in our laboratory. When water was squirted through the screen top of their terrarium, the animals would move over under the drip, open their mouths and catch and swallow the drops. We have no explanation of why frogs adapted to the extremely humid condition of the upper Andes should be the first to be recorded as drinking water.

For the most part amphibians do not live in brackish or salt water, but one species, *Rana cancrivora,* regularly inhabits brackish water and others may be exposed to it on occasion. *Rana cancrivora* is able to maintain osmotic balance with the environment in the same manner as the elasmobranchs; that is, the animal retains enough urea in the body fluids and tissues to bring it into osmotic balance with the saline water in which it lives.

It is well known that several species of *Bufo* (e.g. *B. bufo, B. viridis,* and *B. calamita*) can live up to several months in waters of higher salinity than typical fresh water. *B. viridis* has the ability to adjust to as much as 75 percent sea water. In this species this ability seems to be due to a temporary increase in urea, inorganic ions, and to a lesser extent, amino acids in the plasma.

Osmoregulation in Reptiles

Until rather recently osmotic balance in reptiles had been little investigated. The reptiles convert their nitrogenous wastes into nonsoluble, nontoxic uric acid, which can be temporarily stored without the addition of water. This makes it possible for the embryo to excrete nitogen within the confines of an eggshell. It also means that little water needs to be expended by the adult for the removal of nitrogenous wastes. Indeed, reptile urine is often a semi-solid paste. Until a few decades ago it was generally assumed that in the reptiles water loss took place either through the kidneys or by means of respiratory evaporation and that this was balanced by uptake through the digestive system. It is now known that the reptile skin is not so impermeable as it was previously thought to be. Two-thirds or more of the water loss in most reptiles is through the skin.

Studies of numerous species of turtles, snakes, lizards, and amphisbaenids indicate that the rate of water loss in reptiles is primarily a function of ecology rather than of their size or systematic position. Table 8-3 compares water loss in five species of turtles. These turtles were not fed for a week, then were catheterized to remove urine and accessory bladder water, and were placed in a shaded outdoor cage for five days. It can be readily seen that the percentage of water loss was much greater in the aquatic forms *(Chelydra* and *Sternothaerus)* than in the more terrestrial ones *(Terrapene* and *Clemmys).*

In legless reptiles the difference in the rate of water loss by desert-adapted forms and species from humid tropical forests may be as much as a hundred fold. Variation in the rate of water loss through the skin probably accounts for most of this difference, for in the desert-adapted *Crotalus scutulatus* the water so lost is less than half that lost by respiration.

TABLE 8-3
Water Loss in Several Species of Turtles

Species	Number of individuals	Mean loss in body weight (%)
Terrapene c. carolina	28	3.6
Clemmys guttata	30	8.9
Clemmys insculpta	7	11.1
Chelydra serpentina	4	15.6
Sternothaerus odoratus	24	20.0

Although reptiles lose water readily through the skin, they are apparently unable to absorb it by the same route. Among the snakes no weight gain has been reported for animals kept in moist sand unless their heads were also in contact with moist ground or unless they were able to drink. Most reptiles apparently gain the needed water through their food, though some also drink. The Australian lizard, *Moloch horridus,* is reported to have a network of very narrow, open channels in its skin. If it is placed with its belly in water, the water moves through these channels by capillary attraction to the mouth and is then swallowed.

READINGS AND REFERENCES

Allee, W. C., A. E. Emerson, O. Park, T. Park, and K. P. Schmidt. *Principles of Animal Ecology*. Philadelphia and London: W. B. Saunders, 1949. (A standard, comprehensive work on animal ecology.)

Cowles, R. B., and C. M. Bogert. "A Preliminary Study of the Thermal Requirements of Desert Reptiles." *Bulletin of the American Museum of Natural History,* vol. 83, art. 5, 1944. (In this and other papers, Cowles and Bogert have contributed largely to our understanding of temperature as a factor in the environment of the herptiles.)

Elkan, E. "Mucopolysaccharides in the Anuran Defence Against Desiccation." *Journal of Zoology,* vol. 155, pt. 1, 1968.

Gordon, Malcolm S., et al. *Animal Function: Principles and Adaptations.* New York: Macmillan, 1968.

Moore, John A., ed. *Physiology of the Amphibia.* New York: Academic Press, 1964.

Rahn, Hermann. "Gas Transport from the External Environment to the Cell," in *Development of the Lung* (a Ciba Foundation Symposium). Boston: Little, Brown, 1966.

Szarski, Henryk, "The Structure of the Respiratory Organs in Relation to Body Size in Amphibia." *Evolution,* vol. 18, no. 1, 1964.

9

RELATION TO BIOTIC ENVIRONMENT

Not only do the amphibians and reptiles maintain homeostasis by a more or less constant exchange of such things as heat, water, and respiratory gases, with the physical environment, but they constantly interact with biotic factors as well. These factors include food, competitors, predators, and parasites. It is the combination of physical factors and biotic factors that determines the nature of a habitat.

The environment not only provides the means of existence for the animal —oxygen, water, food, and shelter—but it also imposes restraints upon the animal. It sets the bounds within which the animal must live and largely determines how many of each kind will survive. Every species has an inherent power of reproducing itself, a "biotic potential," in numbers far in excess of the number that actually can or do survive. The existing population of a species at any one time reflects a balance between its biotic potential and the resistance of the environment. When they are in equilibrium, just enough young reach maturity each year to replace the adults that die, and the population thus remains constant. The restraining factors of the environment themselves fluctuate: one year may be too wet, another too dry; one year the food may be abundant, the next may bring famine; another year predators may be numerous, but later there may be few. As a result, populations of animals also fluctuate. If environmental resistance is reduced, the population grows; if environmental resistance increases, the population becomes smaller. Generally these fluctuations cancel each other so that over the years the numbers of a given species remain fairly constant. Drastic or long-term changes in the environment, however, produce permanent changes in populations. European settlers, with their guns and their custom of draining swamps to make farmland, materially changed the environment of the American Alligators and permanently reduced their numbers.

The study of ecology is in part a study of the factors involved in environmental resistance. These factors are many, complex, and interrelated, but they are susceptible to analysis. From such studies certain basic principles have emerged. In 1840 Justus von Liebig first clearly expressed what is now known as Liebig's Law of the Minimum. This simply says that the factor necessary for life that is present in minimum quantity is the one that offers the greatest restraint to a population. For example, if the food of a given species is very scarce in a region, that scarcity determines how many individuals of the species can survive in that region. If food is plentiful but home sites are few, it is the scarcity of available home sites that limits the size of the population.

Liebig's Law has been modified and expanded through the years. Victor E. Shelford has pointed out that it is not always a minimal factor that limits the distribution of an animal, sometimes it is a maximal factor. Too high temperatures, too much water, too much light, too many predators or parasites, restrain a population just as effectively as does too little of something. Thus Shelford expanded the Law of the Minimum into the Law of Tolerance, which says that a species has a range of tolerance for a given environmental factor, bounded on one side by a minimum and on the other side by a maximum, which sets the limits of tolerance for that species. An animal may have a wide range of tolerance for one factor and a narrow range for another. Those animals with the widest ranges for all factors are the ones most likely to be widely distributed. The period of reproduction and development is likely to be the critical one, for many environmental factors are more limiting on eggs, embryos, and larvae than they are on adults.

Since it is relatively simple to set up a controlled experiment in which only a single physical factor is varied, the precise effects of many physical factors can be determined more readily than can the effects of the less easily manipulated biotic factors. But no animal lives in a biotic vacuum. Herptiles are subject to restraints imposed by the living things around them—the predators that prey upon them, the parasites that live at their expense, the competitors that vie with them for the limited necessities of their existence. Many of these restraint, although superficially obvious, are often difficult to study quantitatively.

Interactions between living things are frequently discussed under the heading "biotic relationships." But since "relationship" has also the more restricted meaning of closeness of descent from a common ancestor, it will be less confusing if we call the situations discussed in this section "interactions" and retain the evolutionary implications of the word "relationship."

For convenience we will divide these interactions into "consort interactions," in which there is direct contact between the two organisms, and "nonconsort interactions," in which one form affects another merely by living in the same community.

CONSORT INTERACTIONS

Food

Like all animals, the herptiles are dependent on other living things as sources of food. Usually they are carnivorous. Most amphibians and many reptiles feed on a wide variety of animals. Analysis of the stomach contents of some Greenhouse Frogs *(Eleutherodactylus)* showed that these little frogs had eaten representatives of eight different orders of insects, and also spiders, mites, centipedes, millipedes, and earthworms.

Turtles and crocodilians, which are able to tear off pieces of flesh, can feed on carrion or anything small enough for them to kill. A salamander, *Plethodon jordani,* will seize a large earthworm in its jaws and rotate its body rapidly until the worm breaks. It then swallows the fragment before attacking the remainder. Most herptiles, though, must swallow their food whole, and for them size is an important factor in determining their selection of food. Early in the spring, young grasshoppers constitute a large part of the diet of the Northern Side-blotched Uta *(Uta s. stansburiana),* but in the summer the grasshoppers have grown too large for this small lizard to handle and it then feeds on leaf hoppers.

Many herptiles change their diets as they grow. Young of the Suwanee Turtle *(Chrysemys concinna suwanniensis)* are largely carnivorous, feeding on ant larvae and other small invertebrates, but the herbivorous adults graze on flats of eelgrass near the mouth of the Suwanee River.

Some reptiles have developed highly specialized food habits. Snakes of the genus *Dipsas* in South America prey exclusively on snails, and snails are also important in the diet of the Caiman Lizard *(Dracaena guianensis.)*

Herptiles vary greatly in the amounts that they eat and the frequency with which they take food. Some, such as the toads, eat large meals daily. A toad, indeed, may try to continue eating even when it is stuffed to capacity. A large female American Toad *(Bufo americanus)* ate 152 Mexican Beetles in a day. It has been calculated that from May through September a single toad could eat 22,700 of these destructive insects. This is but one example of an amphibian that benefits man by its feeding habits.

At the other extreme from the toads we find the snakes, which eat much less often. This is probably correlated to a large extent with the snake's ability to distend its jaws and thus swallow large chunks of food. These make a much more lasting meal than the tiny morsels eaten by frogs and lizards.

Many turtles and some lizards depend largely on plants for food. In observing the big Land Iguanas *(Conolophus)* of the Galápagos Islands, Darwin

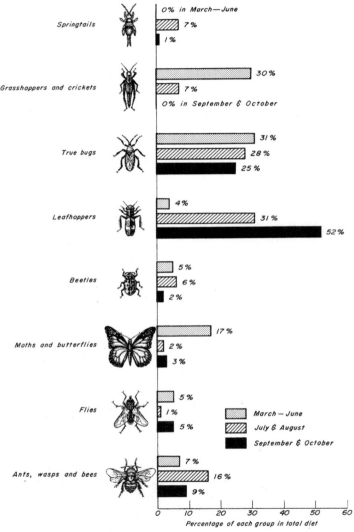

Percentage of each group in total diet

FIGURE 9-1

Seasonal change in food composition of the Northern Side-blotched Uta (*Uta stansburiana*) in Utah. Percentages are those of total number of items represented. Note particularly the increase in the consumption of leafhoppers (mainly, the Beet Leafhopper, *Eutettix tenellus*) as the season progresses and these insects become more abundant.
[Data from George F. Knowlton, after Oliver, 1955.]

FIGURE 9-2
The Caiman Lizard, *Dracaena guianensis*, feeding on a snail.

found to his surprise that "to obtain acacia leaves they crawl up low stunted trees. It is not uncommon to see a pair quietly browsing...on a branch several feet above the ground." He amused himself by tossing a branch among a group of them and watching them attempt to seize and carry it off "like so many hungry dogs with a bone." Shingle-back Lizards of Australia (*Trachysaurus rugosa*) eat berries and toadstools. When these large skinks live close to settled districts, they may become a serious problem. They are particularly destructive to strawberries, and can devastate a crop by eating the unripe berries.

Predation

Although many herptiles are predators, most of them are also small enough to be the prey of many other animals. Tree frogs eat insects and snakes eat frogs. Whether we regard a frog as predator or prey depends upon our point of view.

It is doubtful that anyone could spend much time collecting herptiles without becoming aware of the importance of predation in the lives of amphibians and reptiles. Just one evening spent by a pond where a frog chorus is in full swing will demonstrate that water snakes which feed on the frogs have also been attracted to the same spot. Predation is not usually deleterious to a species though it is, of course, destructive to the individual. Studies on other animal groups have shown that a natural balance in numbers is maintained between the predator and its prey. When predators are removed, a population may become so numerous it destroys or at least seriously depletes its food supply; starvation then reduces the population to a level perhaps lower than it maintained when predators were present. Also, predators usually catch only those animals that are most easily caught: the sick and the unfit. They thus check the spread of diseases and help maintain genetic fitness by eliminating the carriers of deleterious genes. These interactions have been studied among birds and mammals. Undoubtedly they exist among reptiles and amphibians as well, but there is a great need for quantitative studies and critical analyses of the effects of predation on the herptiles.

Studies of two species of Spiny Lizards (*Sceloporus graciosus* and *S. occidentalis*) showed that about 65 percent of the adult population of *S. graciosus* was replaced after an interval of three years, whereas there was 80 percent or more replacement of *S. occidentalis* after only one year; this was attributed in part to differences in predation. In the habitat of *S. graciosus* at the locality of the study, only one predatory snake *Thamnophis sirtalis,* was present, whereas six known or presumed predatory snakes (*Coluber constrictor, Lampropeltis zonata, L. getulus, Pituophis melanoleucus, Thamnophis ordinoides,* and *Crotalus viridis*) lived in the same general habitat as *Sceloporus occidentalis*.

A useful indicator of the extent of predation on lizards is the amount of tail breakage. In the population of the lizard *Cnemidophorus lemniscatus* on the Bay Islands of Honduras, only one-eighth as many individuals have broken tails as in the mainland population. There are few predators on the Bay Islands, but many on the mainland. The Bay Islands population is also much more varied than the mainland population; a variability that may result simply from decreased selection pressure. Many of the variants that are eliminated by predators on the mainland may survive and propagate on the islands.

Symbiosis

This is a special and close association between two kinds of organisms. Literally, they "live together." Symbiotic associations are classified according to whether both members benefit (mutualism), or one member benefits and the other is unaffected (commensalism), or one member benefits and the other is injured (parasitism). The various categories grade into one another so that it is often difficult to say how we should classify a given association. We are often handicapped, too, by lack of knowledge; we know that two forms are closely associated, but we do not know how they affect one another. The categories should be regarded, not as hard and fast divisions, but as convenient groupings for the discussion of a variety of situations.

Mutualism.   Only a few instances of true mutualism have been reported among the herptiles. A species of green alga grows inside the outer membranes of the eggs of the Spotted Salamander *(Ambystoma maculatum)*. This is apparently beneficial to both organisms: eggs that are inhabited by algae produce larger embryos, hatch earlier, and have a lower mortality rate than eggs that do not have algae, and the algae seem to grow more vigorously in eggs containing embryos than in those from which the embryos have been removed. The mechanisms by which plant and embryo affect each other are unknown.

Some aquatic turtles are true mossbacks, their shells being covered with a dense growth of algae. The shell provides a place of attachment for the plants, and the plants help conceal the turtle.

One type of mutualism is "cleaning," in which an organism removes— and eats—deleterious organisms from the body of another. On the Galápagos Islands, large red crabs have been seen crawling over the bodies of Marine Iguanas *(Amblyrhynchus)* and pulling ticks from their skins. Nearly 2,500 years ago, the Greek historian, Herodotus, reported that crocodile birds enter the mouths of basking crocodiles to remove leeches from their gums. The story has never been verified, but if it is true it is undoubtedly the oldest record of this type of symbiosis.

T. W. Kirkpatrick, in his book *Insect Life in the Tropics,* reports a situation that strongly suggests genuine mutualism and that could well repay further investigation:

> In East Africa a "blind snake" (although actually it has vestigial eyes) of the genus *Typhlops* lives with the safari ants, *Dorylus,* and accompanies them when they change their nesting site. I have several times seen one of these short thick snakes near the head of a marching column, with a few ants, apparently path-finders, going ahead and a number riding on its back. The main column carrying the brood followed behind and one got the impression that the snake was being employed as a bulldozer. Nothing is known about the biology of these snakes, though they are probably scavengers in the ants' nest. It is curious that I have never seen more than a single snake accompanying a column of ants, for if they bred in the ants' nest one would expect on occasions to find more than one. An East African native told me that in order to destroy a colony of "siafu" it was only necessary to kill their tame snake, but it seems unlikely that this should be correct.

Commensalism.   This is a one-sided symbiotic interaction from which one member derives some advantage—protection, transportation, or shelter—and the other is not affected. One kind of commensalism is "inquilinism," in which a species lives in a domicile that is constructed and occupied by another. Herptiles may be guests of, or hosts to, many other animals. Perhaps the best known inquilines are the ones that live with the different species of gopher tortoises. Sidewinders *(Crotalus cerastes),* Great Basin Rattlesnakes *(Crotalus viridis),* Spotted Night Snakes *(Hypsiglena torquata),* and Banded Geckos *(Coleonyx variegatus)* have all been found hibernating in dens of the Desert Tortoise *(Gopherus agassizi).* The burrows of the Gopher Tortoise *(Gopherus polyphemus)* are occupied, occasionally, by Diamondback Rattlesnakes *(Crotalus adamanteus)* and, typically, by Gopher Frogs *(Rana capito).* The Tuatara *(Sphenodon)* often makes its home in the burrow of a bird, the Sooty Shearwater.

Many different kinds of protozoans are present in the intestines of herptiles. They are usually considered parasites, but since many seem to do no appreciable injury to the host they should perhaps be classed as commensals. It is even possible that some of them are beneficial.

Parasitism.   Parasitism is a type of symbiotic association in which an organism spends most or all of its life on or in the body of another, derives its food from its host, and frequently damages the tissues of the host. Parasitologists have found the herptiles fruitful sources of parasites, both external and internal. Mites, ticks, leeches, copepods, fly larvae, flukes, tapeworms, threadworms thorny-headed worms, and protozoans have all been reported to be parasites of the herptiles.

Outside that bane of the zoo keeper, the mites, perhaps the best known parasites of reptiles are the flukes (trematodes) that are found in the respiratory and digestive systems of snakes. Most snake autopsies show heavy infections by these trematodes, but such infestations are seldom fatal. Indeed the presence of trematodes does not seem to have much effect on the well-being of the snakes, although studies of the poisonous *Denisonia* and *Notechis* of Australia indicate that parasites may diminish their venom yields.

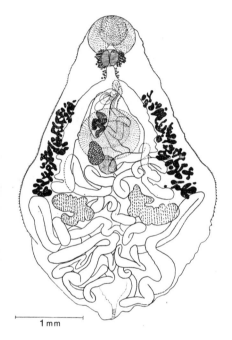

FIGURE 9-3
The lung fluke *Pneumatophilus leidyi*, from a Banded Water Snake, *Natrix sipedon fasciata*. [After Byrd and Denton, *Proc. Helminthol. Soc. Wash. D.C.*, 1937.]

1 mm

There are some interesting correlations between trematode infections and the habits of snakes. Two species of garter snakes *(Thamnophis sirtalis* and *T. butleri)* differ quite pronouncedly; the average specimen of *T. sirtalis* is heavily infected with trematodes, *T. butleri* seldom has them. These parasites pass through the early stages of their development in the body of a snail. If, after they leave the snail host, the larvae are ingested by an amphibian, they undergo further development in its body. If the amphibian is then eaten by a snake, the snake becomes infected and the trematodes reach adult form within it. Since *T. sirtalis* feeds largely on amphibians, it is usually heavily parasitized; since *T. butleri* eats earthworms and leeches rather than amphibians, it is not. The presence of the parasite provides a clue to the diet of the host.

For both an animal species and its parasites to survive, there must be an adjustment between them. If the members of a parasitic species are so numerous and so deadly as to kill all available hosts, then the parasite itself will become extinct. Parasitism therefore tends to evolve toward commensalism. Adjustment between the trematodes and snakes seems to be an adjustment between individuals; that is, the body of an infected snake appears able to withstand the damage done by the parasites. Adjustment may also take place between two populations. In the vicinity of Gainesville, Florida, Carolina Anoles (*Anolis carolinensis*) are parasitized by the larvae of a fly of the family Sarcophagidae. In some years, during May and June, the bodies of dying anoles are found on the ground, their insides eaten away by maggots. As many as 34 maggots have been taken from a single anole. During other years, however, no evidence of such parasitism is found. It is probable that the periods of heavy infection are periods in which the anoles are particularly abundant, and that the numbers of the two populations fluctuate together so that there are never enough flies to reduce the number of lizards below the level necessary to maintain the population.

One of the most interesting studies of amphibian parasitology was the attempt by Dr. Maynard M. Metcalf to correlate the evolution of the ciliate protozoans of the family Opalinidae with the evolution of the frogs they parasitize. Since he found that opalinids are not strictly host-specific, but sometimes transfer from one family to another, studying them did not reveal much about the evolution of the frogs. The main interest of Metcalf's work lies in his method of gathering and combining data from the organisms and their parasites. Since host and parasite evolve together, this method should be useful in future evolutionary studies.

NONCONSORT INTERACTIONS

The interactions discussed above involve direct association between individuals of one species and those of another species. There are, however, other interactions between two or more species of animals or between plants and animals that have a bearing on the well-being of the interacting species although there may be no direct contact between individuals.

Habitat Formation

Plants play a most important role in the general environment of the herptiles. They are a major part of the habitats in which most species live, largely determine the characteristics of these habitats, and have a profound effect on the physical features of an environment. In forested regions plants serve as buffers against extremes of heat and cold, their roots hold moisture in the soil, the water they transpire adds to the humidity of the air, they provide shade, and they contribute to the formation of soils. Rotten logs form protective retreats for many of the herptiles; some terrestrial salamanders may spend most of their lives in the vicinity of a single large log. The leaf mold on the ground provides a habitat for small salamanders, frogs, and snakes. Many herptiles lay their eggs or bear their young in rotten logs, dead stumps, or leaf mold.

Buffer Species

Studies of the feeding habits of otters in Michigan trout streams showed that they ate 25.9 percent forage fish, 25.3 percent amphibians, and 22.7 percent game and pan fish. If the number of forage fish should be reduced, the otters would eat more amphibians and game fish. This is what we call buffering; that is, the amphibians and trout are protected from excessive otter predation by the presence of the forage fish, and the amphibians in turn help relieve predation on the game fish. The number of amphibians present in the area is thus determined in part by the number of forage fish, and in turn helps determine the number of game fish, although there is no direct contact between fishes and amphibians. Since populations of many, if not all, animal species alternate between periods of abundance and scarcity, and since many predators show few specific preferences in food, simply eating what is most readily available (usually the most abundant species), the buffering interaction is likely to be fluctuating and reciprocal. If members of one species are very common in a given year, predators will eat many of them and a less common species will be buffered. The next year, numbers of the first species may be reduced, in part because of the heavy predation. The second species, which was protected the first year, may become the more common one, and the predators will turn their attention to it. A species that was buffered one year may thus be the buffer another year. As far as herptiles are concerned, such interactions are largely theoretical, but it is logical to assume that since they have been reported for other animals, they probably also occur with herptiles. We need long, continued studies of predation, correlated with studies of the relative abundance of prey species, before we can say definitely what does occur.

Mimicry

Among the snakes of South America superb examples of both Batesian and Müllerian mimicry are found. In Batesian mimicry a harmless or edible species evolves (through convergence) an appearance similar to some harmful or distasteful species. Thus the widespread colubrid *Rhinobothryum lentignosum* of the Amazon Basin is superficially strikingly similar to the triad coral snakes of the same area. They all have a pattern consisting of groups of three black rings (triads). The individual rings of a triad are separated by yellow or white rings and the triads are separated by red rings (see Fig. 17–12). The limiting condition of Batesian mimicry is that the mimics must never greatly outnumber the models for if they were to do so predators would be more likely to learn that the pattern signaled something palatable and harmless. Thus if *Rhinobothryum* should become more abundant than the triad coral snakes its mimicry would be valueless. Within the range of the coral snakes similar harmless mimics can be found such as *Atractus latifrons, A. elaps, Procinura aemula,* and *Lampropeltis triangulum doliata.*

On the other hand it is advantageous for the numerous coral snakes in South America to be similar in appearance. This type of evolution, in which two or more harmful or distasteful species come to look alike, is known as Müllerian mimicry. When a number of species adopt the same warning hallmarks the number of individuals that fall victim to predation is neither increased nor decreased as each new generation of predators learns to recognize them, but the total number of victims sacrificed is spread over several different species rather than a single one. The South American coral snakes are probably the best example of Müllerian mimicry to be found among the vertebrates.

Competition

The interactions discussed above mainly involve animals that are not closely related. Competition, on the other hand, is apt to involve members of the same or of very similar species. This is because the more closely two forms are related, the more similar their ecological requirements are apt to be, and the more intense will be the competition between them.

Many closely related species are allopatric (living in different areas). When two closely related species are sympatric (living in the same general area) they almost invariably develop slightly different ecological requirements so that competition between them is reduced. When two species with very similar requirements come to inhabit the same area, such overlapping is usually temporary; one or the other will be eliminated by competition.

Three species of garter snakes occuring in Michigan—*Thamnophis sirtalis,* the Common Garter Snake, *T. sauritus,* the Eastern Ribbon Snake, and *T. butleri,* Butler's Garter Snake—are found together in high concentrations near hibernating sites in the spring and fall. The Common Garter Snake and the Ribbon Snake are about the same size but differ greatly in feeding habits. A recent study showed that amphibians made up 90 percent of the food of the former but only 15 percent of the food of the latter. Earthworms formed 80 percent of the food of the Ribbon Snake and 83 percent of the food of the smaller Butler's Garter Snake. The latter did not eat amphibians but 10 percent of its food was leeches. During the summer months the snakes dispersed.

Butler's Garter Snakes were found most often in grassy areas near water. Ribbon Snakes were also found near water, but in bushy rather than grassy areas. Common Garter Snakes were much less restricted in habitat. They overlapped the other two in distribution and were also found in areas where the others were not usually seen. Thus the two species most similar in food preference, *T. butleri* and *T. sauritus,* live in different minor habitats during the summer months when they are feeding most actively. The species *(T. sirtalis)* that overlaps the other two in habitat distribution differs greatly from both in food requirements. Thus competition between the three is reduced and they are able to live side by side.

Since large animals tend to eat larger food items than small animals, differences in size may reduce competition between two similar species to a point that they are able to coexist in the same habitat. In Jamaica, three closely related species of frogs of the genus *Eleutherodactylus* may occur in the same place; a large *E. pantoni,* a medium-sized *E. gossei,* and a very small *E. andrewsi* may live under the same rock. The difference in size is probably enough to produce a difference in diet so the three are not really competitors.

Difference in size probably helps also to eliminate competition between the young and adults of the same species. During the period of emergence, the young of a species are very numerous as compared to the adults. Competition among the young is very intense and it may be this that compels them to move from the breeding sites. There are indications that new territories are usually invaded by young frogs rather than by adults.

There are probably many other interactions among species of which we know little or nothing. For instance, it has been shown that of the total population of snakes in an area, about half of the individuals belong to about 10 percent of the species and about 10 percent of the individuals belong to half of the species. This holds true in Panama, with its numerous and diverse snake fauna, and in places like Pennsylvania and California where fewer species and fewer individuals are found. Why this is so is not known, but it is hardly likely that these ratios, occurring as they do in such distant geographic areas, are coincidental.

READINGS AND REFERENCES

Allee, W. C., A. E. Emerson, O. Park, T. Park, and K. P. Schmidt. *Principles of Animal Ecology.* Philadelphia and London: W. B. Saunders, 1949. (A standard, comprehensive work on animal ecology.)

Angel, F. *Vie et Moeurs des Amphibiens.* Paris: Payot, 1947.

———. *Vie et Moeurs des Serpentes.* Paris: Payot, 1950.

Metcalf, M. M. "The Origin and Distribution of the Anura." *The American Naturalist,* vol. 57, Sept.–Oct., 1923.

———. "Further Studies on the Opalinid Ciliate Infusorians and Their Hosts." *Proceedings of the United States National Museum,* vol. 87, no. 3077, 1940.

Noble, G. K. *The Biology of the Amphibia.* New York: McGraw-Hill, 1931. Reprinted by Dover Publications, 1954.

Oliver, James A. *North American Amphibians and Reptiles.* Princeton, N.J.: Van Nostrand, 1955.

Smith, M. A. *The British Amphibians and Reptiles.* Rev. ed. London: Collins, 1954.

10

BEHAVIOR

If we approach a pond on whose banks a number of frogs are sitting quietly, we are greeted by a series of splashes as the frogs jump into the water and swim to hide under the detritus at the bottom. They have been stimulated by the sight or sound of our approach and have responded with an escape reaction. In so doing they have exemplified one definition that is sometimes given of behavior—that it is a response to a stimulus. A stimulus in turn is defined as some change in the external or internal environment to which the organism responds. The definition of behavior is thus really circular. It is also too broad. A neotenic salamander may respond to an increase in the amount of iodine in the water by metamorphosing, but this is not usually regarded as behavior. Instead of attempting to give a precise definition, we shall list some of the characteristics of those responses that are considered behavioral.

CHARACTERISTICS OF BEHAVIOR

Behavior is selective. Out of a number of possible responses to a stimulus, an animal makes one. The frogs at the pool could have remained in place, or they could have turned to face us, or perhaps they could have made a threatening display. This is not to say that a frog deliberately chooses which response to make. Its response is determined by a complex of internal and external factors beyond its control.

Behavior is usually reversible. If we sit quietly at the pool for awhile, one by one the frogs will appear at the surface, swim to the bank, and resume their former positions.

Behavior can be more or less modified by experience. If we approach the pool several times a day for several days, at least some of the frogs will no longer respond to our approach by jumping into the water.

THE STUDY OF BEHAVIOR

There are three approaches used to study behavior. The neurophysiologist is concerned with the neuronal and hormonal processes that are the basis of the stimulus-response mechanism. The animal psychologist is interested in the behavior of an animal under controlled conditions, with special emphasis on the modifications of behavior through such mechanisms as the conditioned reflex and trial-and-error learning. The ethologist, who studies the actual behavior of the animal under conditions that are as natural as possible, is primarily concerned with the innate behavior patterns of the animal, and the adaptive and evolutionary significance of those patterns. In this chapter we shall concentrate on the ethological approach.

DETERMINANTS OF BEHAVIOR

Behavior is controlled by the animal's genetic constitution and its physiological condition.

Genetic Factors

We have no information on the genetics of behavior patterns in the herptiles. That these patterns are inherited is indicated by the fact that they are innate. They do not have to be learned and they are characteristic of the species to which the animal belongs. Frequently they are species-specific, differing in a characteristic way from similar behavior patterns in related species. As is typical of iguanids, a male Fringe-toed Lizard (genus *Uma*) will put on a challenging display when approached by another lizard. A part of this display is a series of push-ups in which the front legs are alternately extended and flexed so that the fore part of the body moves up and down. The series in the Mojave Fringe-toed Lizard *(U. scoparia)* consists of two (sometimes three) rapid push-ups, a pause with the body held in the low position, another push-up, another pause, and then the lizard rises to half height. The sequence in *U. notata* begins with two push-ups that are not so rapid as those of

U. scoparia. In the next push-up, the body of the lizard returns only to half height rather than to the low position. This is followed by a slight rise, a slight dip, and then the lizard rises to full height (see Fig. 10-1).

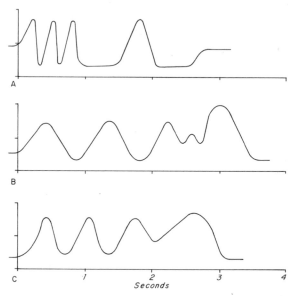

FIGURE 10-1

Push-up sequences in Fringe-toed Lizards. The height of the line indicates the relative elevation of the forepart of the lizard's body during a single sequence. (*A*) *Uma scoparia*; (*B*) *U. notata*; (*C*) Some individuals of *U. n. inornata*. [Redrawn from Carpenter, *Copeia*, 1963.]

Since behavioral characteristics are genetically determined and since they frequently show differences between closely related species, we should expect to find intraspecific differences as well. Everyone who has closely observed living reptiles and amphibians is aware that individual differences in behavior do exist. Some members of one race of Fringe-toed Lizards *(Uma notata inornata)* omit the slight rise and dip during the push-up sequence of the challenging display. Whether there is a genetic basis for this difference is unknown.

Physiological Factors

The physiological state of an animal is important in determining how it behaves. A hungry snake pays attention to a mouse in its cage, a snake that has recently fed may ignore the mouse. It has been shown that the ability of a Desert Iguana *(Dipsosaurus dorsalis)* to learn a maze increases as the lizard's body temperature approaches the eccritic temperature.

Hormones.   The relationship between hormones and behavior is most obvious in the differences in activity between the nonbreeding and breeding seasons. Combat between rival males is characteristic of many courting reptiles. Injections of testosterone have been found to elicit combat behavior in nonbreeding males of the Texas Tortoise *(Gopherus berlandieri)* of the Painted Turtle *(Chrysemys picta)* and of lizards of the *Sceloporus torquatus* group.

Endogenous Rhythms.   Many kinds of animals are known to show rhythmic changes in activity; these are known as endogenous rhythms. These changes are correlated with physiological changes in the animal and are timed to correspond to changes in the environment, to the alternation of day and night, to the phases of the moon, to the seasons, and to the tides. Nevertheless, they are not dependent on external stimuli. Lizards of the genus *Lacerta* hatched from eggs kept under constant conditions and reared in continuous darkness still showed periods of rest and activity that conformed to a twenty-four hour cycle (see Fig. 10-2). Such a daily rhythm is known as a circadian rhytm. A great many species of herptiles are known to be either nocturnal or diurnal, but there is need for further study to determine whether these rhythms of activity are simply responses to variations in such external factors as light, heat, and humidity or whether they are truly endogenous.

Rhythms related to phases of the moon are also known for a number of species of animals. On Java, the toad, *Bufo melanostictus*, breeds throughout the year though the height of the breeding season corresponds to the onset of the northwest monsoon rains in November and December. A daily record was kept of the number of pairs of amplexing toads in a breeding area throughout a year. During periods of the waxing moon, 741 pairs were observed in amplexus; during the waning moon only 431 pairs were observed. Whether this represents a true endogenous rhythm has not been determined.

Seasonal changes in activity are marked in many herptiles. Hibernation is an inherent, regular, and prolonged period of inactivity during the winter, and is in contrast to retraherence, which is a temporary retreat from adverse weather conditions. In northern Florida, Broad-headed Skinks *(Eumeces laticeps)* hibernate early in October and do not emerge until March even though in this region there may be many days during the winter when the temperature is well within the activity range of these lizards. Carolina Anoles *(Anolis carolinensis)* in the same region are active on warm days throughout the winter and remain in shelter only during cold spells. British newts *(Triturus)* kept under constant temperature in a laboratory failed to hiber-

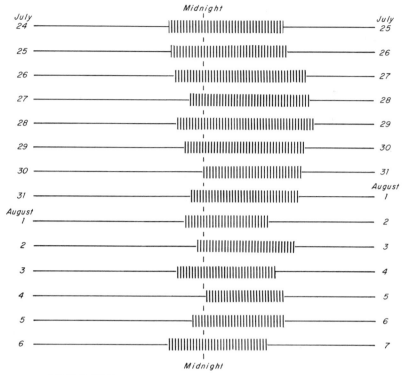

FIGURE 10-2
Periods of activity in a lizard, *Lacerta,* incubated and reared
in total darkness. [After Hoffman, *Z. Vergleich. Physiol.,* 1959.]

nate. These facts suggest that some herptiles show true, endogenous hibernation and others simply undergo more or less prolonged periods of retraherence in response to exogenous factors.

Aestivation is a regular, inherent retreat from hot, dry climatic conditions.
The Sardinian Cave Salamander *(Hydromantes genei)* has been reported to
aestivate in summer even in the laboratory where humid conditions were
presumably maintained. Captive Australian Long-necked Turtles *(Chelodina
longicollis)* tucked themselves away for weeks without feeding in the summer even though their ponds were kept filled with water. Here again there
is a suggestion that endogenous rhythms may be involved.

ELEMENTS OF BEHAVIOR

An animal may be aroused to activity by a complex of internal factors that
give rise to a motivational state, or drive: a hungry snake searches for food.

Or, the activity may result from an external stimulus: the approach of a potential predator elicits escape-behavior in the frogs around the pond. Frequently, both the motivational state and an external stimulus are involved: a male frog at the breeding pond attempts to clasp any approaching object of the apropriate size. Outside the breeding season, his motivational state is different and he might either ignore the object or attempt to avoid it.

Fixed Action Patterns

The motor components of behavior consist of a series of fixed action patterns (FAPs). A fixed action pattern is a highly coordinated motor activity that an animal can perform without previous exercise or learning. It is innate and the sequence of motor elements is constant. The lizard push-up movement is a fixed action pattern. Such an activity is neither a simple reflex nor a chain of reflexes. With a reflex, the same stimulus always elicits the same response. Whether or not an animal performs a given FAP depends on several factors, including its motivational state, the length of time since the action was last performed, and the animal's previous experience.

When a Hognose Snake *(Heterodon)* is threatened, it rears up, flattens and spreads its neck, hisses, and often lunges with closed mouth at the intruder. If this threatening display does not drive away the supposed attacker, the snake "plays dead": it flops on its back, lolls its tongue, and goes limp. Both the threat display and the death-feigning comprise sequences of FAPs. Most Hognose Snakes kept in captivity soon stop responding to teasing with this particular behavioral pattern. The stimulus that once evoked the response no longer does so.

Sign Stimuli

A given fixed action pattern is usually performed in response to a specific stimulus known as a sign stimulus, or releaser. A sign stimulus is typically very simple. An animal reacts usually to a single aspect of a situation rather than to the total situation. During the breeding season a male lizard of the *Sceloporus torquatus* group will challenge and then attack another male engaged in courtship activities. A courting male of this genus approaching a female gives a characteristic series of short, fast, bobbing motions of the head. When an investigator attached a string to the head of a carved wooden model and bobbed it in this way, it elicited a challenge display from a male. Rubber models, dead females, and live individuals of many other species were also effective as long as they appeared to give the courting bob. The characteristic motion of the head is a sign stimulus that produces challenge and attack in these lizards.

Simultaneous Arousal

Normally, an animal responds to only one motivational state at a time. The arousal of one behavior pattern inhibits the appearance of others. Sometimes, though, two conflicting drives are operative at the same time. Several types of behavior are associated with simultaneous arousal.

Conflict Behavior.   In conflict behavior, both drives are shown at the same time. The Chuckwalla *(Sauromalus obesus),* when exposed to high temperature, pants to dissipate excess body heat. This entails fast, shallow breathing with the mouth gaped and the blood-engorged tongue extruded. The lizard escapes predators by moving head first into a rocky crevice and inflating its lungs. In lung inflation, the mouth is closed each time the lizard swallows a mouthful of air and then the glottis closes to hold the entrapped air in the lungs. The legs and feet of a Chuckwalla were taped to a table and the lizard was heated by an infrared lamp to induce panting. Then the investigator pinched the base of its tail. Three types of response were noted: the lizard stopped panting and inflated; or it started to inflate then deflated and resumed panting; or it stopped panting and inflated, but continued to gape. When the lungs were fully inflated, the tongue extruded from the gaping mouth but the glottis was tightly closed. This is characteristic conflict behavior: the simultaneous arousal of the drives for thermoregulation and for escape resulted in a behavior pattern that showed elements of both.

Displacement Activity and Ritualization.   Sometimes the simultaneous arousal of two conflicting drives results in a motor activity that is not characteristic of either. During aggression an animal is frequently faced with the conflicting drives of attack and flight and displacement activities are common in such situations.

Testudine tortoises investigate potential articles of food by olfactory movements that are species specific. The head is either thrust forward or swung laterally from side to side. During the breeding season, males are aggressive. A male Land Tortoise *(Geochelone carbonaria)* will challenge another male by moving its head in an exaggerated, formalized version of the typical olfactory FAP. The challenged male responds with a similar movement. It is probable that the FAP associated with olfaction first became part of the challenge sequence as a displacement activity elicited by simultaneous arousal during aggression. In these tortoises it has become incorporated into the sequence of aggressive behavior and now acts as a sign stimulus that evokes a response from another tortoise. When a fixed action pattern that is part of one behavioral sequence becomes a part of another sequence in which it serves as a releaser it is said to be ritualized. Frequently the ritualization of a displacement activity leads to the evolution of structures or colors that are emphasized or displayed by the activity and so reinforce the signal function. It is probable that the prominent throat fans of lizards of the genus *Anolis* arose in this way.

SEQUENCES OF BEHAVIOR

Most behavior of an animal is organized into rather definite sequences of actions. Typically the sequence consists of appetitive behavior, taxis, and the consummatory act.

Appetitive Behavior

Appetitive behavior is searching behavior that is elicited by a motivational state. The animal usually moves away from its resting place. It is sensitized by its motivational state to respond to certain stimuli to which it does not respond at other times; it can be said to be searching for the stimulus. Appetitive behavior is usually less stereotyped than other patterns of behavior. The animal may wander more or less at random as male snakes seem to do during the breeding season. On the other hand, newly emerged hatchling sea turtles *(Dermochelys* and *Caretta)* have been reported to perform a highly stereotyped circling movement that is apparently appetitive.

Taxis

When, as a result of appetitive behavior, the animal encounters the stimulus to which it is sensitized, it orients itself in relation to the stimulus. Such an orientation movement is called a taxis. The hungry snake that encounters a prey animal arranges its body so that it is in a position to seize the prey. The hatchling turtles orient toward the broad, open expanse of light that indicates the position of the sea.

Consummatory Act

When a female turtle who has been searching for a nesting site finds the appropriate spot, she digs a hole, deposits her eggs, and buries them. The hunting snake that finds a frog catches and eats it. The foregoing are examples of consummatory acts. Such an act consists of a sequence of fixed action patterns and is the least modifiable part of an animal's behavior. A consummatory act typically ends a behavioral sequence because it changes the motivational state of the animal. The turtle, once she has deposited her eggs, returns to the water.

Directional Travel

The original appetitive behavior and orientation of an animal may not lead directly to a consummatory act; they may instead be followed by directional travel. During the breeding season, adult amphibians move to breeding ponds, and many species return year after year to the same ponds ignoring other bodies of water that are apparently just as suitable and accessible. In England a colony of toads (probably *Bufo bufo*) that was under observation for ten years regularly migrated three kilometers between the breeding pond at which they spent the summers and a sand pit in which they hibernated. Individuals of the California Red-bellied Newt *(Taricha rivularis)* show a strong tendency to return year after year to the same breeding site; of 262 males marked from a single pool in 1953, 85 percent were retaken in subsequent years, almost all from the same pool. (The newts spend the nonbreeding season on mountain slopes that are generally far above the stream.)

A number of species of amphibians (e.g., *Rana catesbeiana, Ascaphus truei, Acris gryllus, Taricha torosa*) have been shown to be able to use celestial cues for guidance of directional movements. For example, Cricket Frogs *(Acris gryllus)* that were removed from their home shore and released in the center of a test pen from which only the sky was visible, were able to get their bearings and move in a direction perpendicular to the home shore both during the day and on moonlit nights. It was not necessary for the frogs to see the sun, but when the sky was completely overcast, their movement was random (see Fig. 10-3). The frogs apparently have both a sun-compass sense and an internal clock that allows them to compensate for the changes in position of the sun during the day.

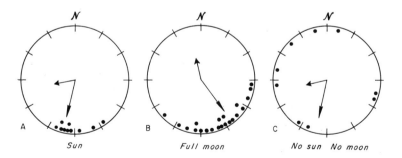

FIGURE 10-3
Orientation in Cricket Frogs, *Acris gryllus*. The long arrow indicates the line parallel to the home shore of the frogs. The short arrow indicates the direction of the home pond. [Redrawn from Ferguson, Landreth, and McKeown, *Anim. Behav.*, 1967.]

Celestial orientation is widespread among animals and it has been suggested that it is the basic mechanism for directional travel among the herptiles. On the other hand, some species at least must use other cues. This is demonstrated by the Spotted Salamander *(Ambystoma maculatum)* which moves to breeding ponds on rainy or cloudy nights.

The most spectacular navigational feats among the reptiles are performed by sea turtles in their migrations to their breeding beaches. There is some evidence that the Green Turtle *(Chelonia mydas)* migrates between feeding areas on the coast of Brazil and the island of Ascension—a distance of 2253 kilometers.

Directional travel may be considered as a special form of appetitive behavior. As with other forms of this type of behavior, it is terminated when the animal encounters a stimulus that "turns off" the migratory phase and initiates a new pattern of behavior.

SOCIAL BEHAVIOR

Members of a population in a given area are frequently found not to be scattered at random throughout the area but to be gathered together in more or less compact groups or colonies. If the animals happen to be in the same place because they are responding to the presence of food or to some physical factor in the environment, the group is called an aggregation. When the animals congregate and/or remain together as a result of the way in which they react to one another, they are considered to form a social group. Many congregations of amphibians and reptiles have been recorded. The breeding choruses of frogs and "balls" of hibernating snakes are perhaps the best known. The tadpoles in a given pond are sometimes found to be gathered in a dense mass rather than scattered at random. This has been most often reported in the Spadefoot Toads *(Scaphiopus)* but has also been noted occasionally in other forms. The Marine Iguanas of the Galápagos Islands *(Amblyrhynchus)* form close-packed colonies. Thirty-one specimens of a gekkonid lizard *(Tarantola annularis)* were found within a crack in a boulder in a very arid region. The plane space available as a resting surface was about three square feet. An inch-thick layer of fecal droppings at the bottom of the crack indicated that this congregation was of long standing.

It is frequently assumed that such concentrations are aggregative rather than social. Thus, it was suggested that the geckos congregated in the crack to sleep during the day because it offered the only available shelter in a very barren area against intense solar radiation and perhaps enemies. Almost always when such assemblages are investigated in detail, though, it is found that the individual derives a benefit from the presence of the other animals. It has been shown, for example, that when salamanders of the genus *Amby-*

stoma are exposed to dehydration, they tend to congregate and that the rate of water loss for a single animal in a container is much greater than when two or more animals are in the container together. It is possible that the presence of so many geckos in the crack raised the humidity of the air and so reduced the danger of desiccation. Under such circumstances, natural selection would favor the development of true social groups. It is probable that many of the so-called aggregations of amphibians and reptiles have a social basis.

Ringneck Snakes *(Diadophis punctatus)* are very gregarious. In the field, half a dozen or more may be found under one small log or rock. A series of these snakes were tested in a laboratory arena provided with ten flat discs as cover objects. Physical and chemical conditions under and around all the discs were the same. The tendency of the snakes to congregate is shown in Table 10-1. The snakes seemed to follow each other or to trail each other across the sand and showed a decided preference for substrates that had been occupied by other Ringnecks. These snakes are thus truly social.

TABLE 10-1

Dispersal of 40 Male Ringneck Snakes in
an Arena Containing Dry Sand and 10 Twelve-inch Discs

Days after release	Number of individuals										Exposed	Hidden
	Under disc											
	1	*2*	*3*	*4*	*5*	*6*	*7*	*8*	*9*	*10*		
2	1	1	1	5	6	4	0	5	1	2	14	0
3	0	2	1	2	4	7	0	6	0	2	16	0
4	1	2	0	1	4	14	0	6	0	4	8	0
5	1	2	2	0	4	17	0	4	0	2	8	0
6	1	1	1	2	4	12	0	6	1	2	10	0
7	0	2	0	0	3	24	1	6	0	0	4	0

Source: Data from Dundee and Miller, 1968.

Advantages of Group Life

We have already alluded to the increased chance of finding a mate as one of the advantages of group life. But there are many others.

Environmental Alteration. The presence of a number of individuals in one spot may change conditions in a way that is beneficial to the animals. Elevation of humidity and a consequent reduction in the rate of dehydration is but one example.

In a cluster of about 180 tadpoles of the Pacific Treefrog *(Hyla regilla)* the animals were oriented in such a way that the tails of about three-fourths of them were pointed toward the sun. This presumably resulted in the maximum exposure of their dark dorsal surfaces to the sun's rays and increased the amount of heat absorbed by the mass of tadpoles. The temperature of the water in the part of the pool where the tadpoles were congregated was higher than in other parts of the pool. Increase in temperature increases the metabolic rate and so speeds metamorphosis.

Spadefoot tadpoles develop in temporary, shallow ponds where there is much danger that the water will evaporate before the toads have a chance to metamorphose. In rapidly drying ponds, 'scooping' congregations have been observed. The tadpoles gather in a group on the bottom and by the lashing of their tails make or deepen a depression in which water may remain long enough for them to metamorphose or at least to survive until the pool is filled again by a fresh rain.

Protection from Predation.   Although a group of animals is more apt to attract the attention of a predator than is a solitary individual, it is probable that the protection provided by the group usually outweighs this disadvantage. The escape behavior of one individual that has been alarmed by an approaching predator may serve as a sign stimulus to elicit the escape reaction in others. Anyone who has approached a group of turtles basking on a log or along the bank of a stream knows how rapidly all the animals plop into the water and disappear. This is usually interpreted as indicating that they all become aware of an approaching predator at nearly the same time, but in light of what is known of similar responses in other vertebrate groups it is probable that the escape behavior of one turtle serves as a warning signal to the others.

That a predator is more apt to attack an isolated individual than a large congregation constitutes selection pressure toward the tendency of the animals to remain close to each other. Fer example, predacious beetle larvae have been reported to avoid entering social groups of spadefoot tadpoles, though they readily attack scattered individuals.

Social Facilitation.   In social facilitation a stimulus provided by the activity of one member of the group elicits similar behavior in other members of the group. Laboratory tests of male Spring Peepers *(Hyla crucifer)* showed that over 90 percent of the frogs called while a tape recording of calling peepers was being played, but less than 5 percent called when the tape recorder was not played. It seems clear that the buildup of the large breeding choruses characteristic of this species depends on the tendency of the frogs to respond by calling to the sound of the voices of other individuals.

Social facilitation of feeding behavior has been reported for tadpoles of the Plains Spadefoot *(Scaphiopus bombifrons)*. When different numbers of tadpoles were placed in culture dishes and food was added, the greatest feeding activity was noticed in the dish with the largest number of individuals. Social feeding groups of tadpoles have been observed in nature in this and other species of spadefoots. Frequently the activity of the tadpoles is such as to stir the bottom detritus so that more bits of organic material are made available to all members of the group.

Social Organization

Animals that live in social groups usually develop special behavioral mechanisms that tend to minimize aggressive reactions among members of the group. These lead to the development of patterns of social organization. Two types of social organization found among many diverse vertebrates are territoriality and social dominance.

Territoriality.   An animal does not wander completely at random but remains within a home area, a restricted stretch of familiar territory within which it conducts its normal activities. Animals that undertake extensive migrations often have more than one home area. Thus a male frog may have one home area away from the breeding pond during part of the year and, when he arrives at the pond during the breeding season, he establishes a new home area along the bank. Turtles and snakes frequently have extensive home areas, but those of lizards are usually small. When a Syrian Fringe-toed Lizard *(Acanthodactylus tristrami)* was repeatedly frightened away from its burrow, it seemed to be thoroughly familiar with the terrain for about 40 square meters around the entrance to its burrow. If it was driven to the limits of this area, it became uncertain, stopped, and changed direction (see Fig. 10-4). This suggests that this animal had a very small home area.

Within the home area the animal usually has a home site to which it regularly returns when not foraging for food or carrying on other activities. The home may be a burrow, a particular leaf on a tree (as for some tree frogs), or simply a restricted area, such as a strip along the bank of a stream, within the home area. When an animal defends the vicinity of its home site against intruders of the same species, it is said to show territoriality; the defended area is the territory. This type of behavior occurs most often among males, who frequently drive away other males during the breeding season, but sometimes females, and even young, show territoriality.

Defense of territory at the breeding site has been observed in males of a number of species of frogs (e.g., *Rana clamitans, Hyla faber, Leptodactylus*

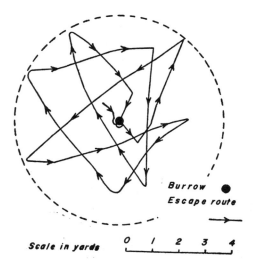

Burrow ●

Escape route

Scale in yards

FIGURE 10-4
The route taken by a Syrian
Fringe-toed Lizard, *Acanthodactylus
tristrami*, when attempts were made
to force it away from its burrow.
[After Riney, *Copeia*, 1963.]

insularum, Dendrobates galindoi). The frog may simply warn the intruder by a threatening display but sometimes shoving or wrestling bouts develop. Territoriality is well developed in the South American frog, *Phyllobates trinitatus.* One frog will hop on top of another in an effort to drive it away. Adult females are the most aggressive, but even very young frogs may have territories. The territories are feeding, not breeding, areas.

Defense of territory is common in lizards. The Marine Iguanas *(Amblyrhynchus cristatus)* of the Galápagos Islands live along the coast, rarely going more than 12 to 15 meters inland. Along the shores, the young and females mass together, oftentimes piled atop one another two or three deep, all within about 9 meters of the water's edge. As many as 76 have been observed within a space of 9 square meters. The older males, which are much larger than the females, take positions between the massed females and young, and the water. They are generally 1.5 to 3 meters apart, and each keeps to his own sunning territory. Every trespass by one male into the sunning terrain of another is the occasion for a fight. The contestants butt one another, each endeavoring to get his horny, knobby head beneath his opponent's chin. On some of the islands, the iguanas are found assorted in family groups, composed of one large male with two to four females. The families are separated from other groups by 18 to 45 meters of shoreline.

Usually the possessor of a territory has the advantage in an aggressive encounter and is able to drive away the intruder. Territory is more often maintained by display than by fighting. The display is usually highly stereotyped and ritualized. A male Bullfrog *(Rana catesbeiana)* in his calling territory floats high in the water with his head raised to display his yellow throat and calls frequently. If another frog in the high position approaches, the resident frog challenges by giving a short, staccato, hiccupping call and swimming a short way toward the intruder. If the second frog continues to call and/or approach, the first male repeats the challenge. In 58 of 79 such encounters observed, the incident terminated when one of the frogs left the area. The rest culminated in pushing and wrestling bouts. The frog that out-pushed its opponent remained high in the water and continued to call; the other adopted a low position by deflating his lungs and swam away. Frogs that had not succeeded in establishing a territory consistently maintained a low position in the water and were not challenged or attacked. High position acts as a sign stimulus that evokes aggression and low position apparently inhibits attack.

Territoriality tends to space the animals and thus maximizes the use of the area's resources, while at the same time, the animals retain the advantages of the group. For the Bullfrog, warning against predators and social facilitation in calling may both be important advantages of sociality. The females are attracted to large choruses but apparently not to individuals calling alone. As with many other animals, males that fail to establish territories do not mate. These are presumably the weaker, less well-adapted members of the population. Territoriality thus tends to maintain the genetic fitness of the species.

Social Dominance.   Many animal societies are organized into dominance hierarchies. If one animal is more aggressive, more successful in attacking others, an animal below him in the hierarchy will seek to avoid aggressive encounters with him by fleeing or by adopting a submissive posture that inhibits attack. The second animal may in turn be dominant to another animal lower in the hierarchy. As with territoriality, dominance, once established, is more often maintained by ritualized display than by actual fighting.

Dominance may be combined with territoriality. Of nine male Black Iguanas *(Ctenosaura pectinata)* that occupied a stretch along an old wall, the dominant or tyrant male occupied the center position. He could and did invade the territories of all the others. The one on his left and the one on his right were lower in the hierarchy. They could invade the territories of those to their left or right, respectively, but not the territory of the tyrant. The three on each end of this chain apparently had to tolerate the invasions of the tyrant or the subtyrants, but themselves could not invade any other territory.

Many breeding male frogs show a type of dominance hierarchy in calling sequence. Choruses of the Southern Spring Peeper *(Hyla crucifer bartramiana)* are made of many individual trios. A trio is initiated by a single individual who calls a varying number of times. Soon he is answered by another and the two then call in sequence. Then a third member joins to form a trio. Figure 10-5 is a musical rendition of a trio of Spring Peepers.

FIGURE 10-5

A musical rendition of the formation of a calling trio of the Spring Peeper, *Hyla crucifer*. [After Goin, *Quart. J. Florida Acad. Sci.*, 1948 (1949).]

Similar duets, trios, and quartets have been reported for a number of other species. Apparently the same individual initiates a trio each time. He is considered the dominant male. The Foam-building Frog *(Engystomops pustulosus)* has such a choral structure. During the night, males drop from the call sequence in reverse order, that is, the low frog in a sequence leaves first. Females that enter the pond in the pauses between sequences of calls move to the dominant male as soon as he initiates a new sequence. Thus he has a greater chance of reproductive success both because he calls for a longer period of time and because he has a greater chance of attracting females.

Ethological research on vertebrates has been conducted mostly on birds and fish, and to a lesser extent on mammals. Reptiles and amphibians have been almost ignored in this research; but the many examples given in this chapter indicate that the principles of behavior formulated on the other groups apply to the herptiles as well and that these animals deserve further ethological study.

READINGS AND REFERENCES

Duellman, W. E. "Social Organization in the Mating Calls of Some Neotropical Amphibians." *The American Midland Naturalist,* vol. 77, no. 1, 1967.

Dundee, H. A., and M. C. Miller III. "Aggregative Behavior and Habitat Conditioning by the Prairie Ringneck Snake, *Diadophis punctatus arnyi." Tulane Studies in Zoology and Botany,* vol. 15, no. 2, 1968.

Eibl-Eibesfeldt, I. *Ethology: The Biology of Behavior.* New York: Holt, Rinehart and Winston, 1970. (Written by a leading ethologist, this is by far the best text available.)

Gorman, G. C. "The Relationships of *Anolis* of the *roquet* species group (Sauria: Iguanidae), III: Comparative Study of Display Behavior." *Breviora,* no. 284, 1968.

Hamilton, W. J., III, and P. R. Marler. *Mechanisms of Animal Behavior.* New York: John Wiley and Sons, 1966.

McGill, T. E., ed. *Readings in Animal Behavior.* New York: Holt, Rinehart and Winston, 1965. (See especially the papers by E. H. Hess and W. H. Thorpe for a background of ethological concepts.)

11

MECHANISMS OF
SPECIATION

The early, classical studies of evolution were concerned with the vast sweep of change through ages of time. At the lowest level, they dealt with the shift from species to species, genus to genus, over perhaps a million years, and paid little attention to the primary sources of the variations that are the building blocks of evolution. On the other hand, early genetic studies stressed the individual and the changes in its genes and chromosomes that make it different from other individuals of the same kind. Many geneticists were little concerned with fitting these changes into the broad canvas painted by the evolutionists, and there was little meeting of minds between the two schools. Now the concerns of the evolutionists and the geneticists are drawing closer together. The modern field of population genetics is concerned with the ways in which mutations are spread through animal populations to make them different from other populations of the same species, and the accumulation of large series of some fossil forms has made possible the application of genetic principles to at least a few extinct populations. To understand evolution we must study both the broad outlines of long-term changes in form and the minute details of changes within the nucleus of the cell.

SOURCES OF VARIATION

Both chromosomal and gene mutations have played a part in the evolution of amphibians and reptiles.

Evolution of Karyotypes

Recently, great strides have been made in the study of the chromosomes of amphibians and reptiles. The karyotypes (number, size, and shape of the chromosomes) of many species belonging to various families have been described in detail and emerging from these descriptions is a pattern of the evolution of the karyotypes. One hypothesis that has been advanced is that the primitive ancestral stocks had chromosomes of two basic types: large (macro) and small (micro): and that the primitive condition was to have acrocentric chromosomes, in which the centromere—the point to which the spindle fiber attaches during cell division—is at one end.

Fusion of Chromosomes. Evolution of chromosomes in both the reptiles and the amphibians seems to have been along the line of fusion of acro-centrics in the region of the centromeres so that, in general, the more modern, specialized forms not only have smaller numbers of chromosomes, but tend to have more V-shaped, metacentric ones, in which the centromere is at the point of the V, and fewer I-shaped acrocentric ones. Furthermore, it appears that in the main macroacrocentrics have fused to form macro-metacentrics and that microacrocentrics have fused to form micrometa-centrics rather than that microacrocentrics and macroacrocentrics have fused. Pericentric inversions, sometimes followed by centric fission, have probably also occurred but these mechanisms seem to have been less impor-tant than centric fusion in the evolution of karyotypes of the herptiles.

Curiously, though, the amphibians seem to have undergone more fusion than have the reptiles. Only among the more primitive amphibian families—Hynobiidae, Cryptobranchidae, Ascaphidae, Discoglossidae, and Pipidae—do we find acrocentrics present at all. Fusion of acrocentrics to form meta-centrics has not proceeded apace in the reptiles; acrocentrics are present in most species and metacentrics are usually lacking in members of the family Lacertidae. The state of fusion is also more prone to individual variation in the reptiles. Thus some Oregon Alligator lizards, *Gerrhonotus multicari-natus scincicaudus,* have eighteen acrocentrics and two metacentrics, but others of the same race have twenty acrocentrics and only one metacentric.

The karyotype pattern of salamanders differs from that of frogs and rep-tiles. The number of chromosome pieces is greater in the primitive sala-manders than it is in the primitive frogs and reptiles and the chromosomes are not so readily divisible into macrochromosomes and microchromosomes.

Although the same method of chromosome evolution by centric fusion seems to have occurred in this group as in the others, additional mechanisms must also have been at work. It has been suggested that a series of losses and fusions of small chromosome fragments has been more important in the evolution of salamanders than in that of other herptiles. There seems to be no close cytological connection between the Caudata and the Anura.

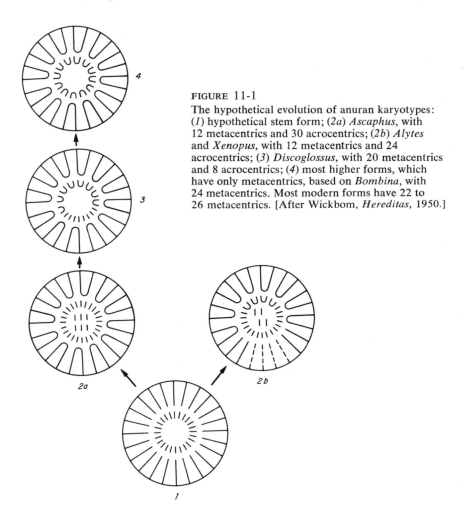

FIGURE 11-1
The hypothetical evolution of anuran karyotypes: (1) hypothetical stem form; (2a) *Ascaphus*, with 12 metacentrics and 30 acrocentrics; (2b) *Alytes* and *Xenopus*, with 12 metacentrics and 24 acrocentrics; (3) *Discoglossus*, with 20 metacentrics and 8 acrocentrics; (4) most higher forms, which have only metacentrics, based on *Bombina*, with 24 metacentrics. Most modern forms have 22 to 26 metacentrics. [After Wickbom, *Hereditas*, 1950.]

The hypothetical evolution of frog karyotypes is illustrated by the diagram (see Fig. 11-1) which shows how the fusion of parts decreases the number of both microchromosomes and macrochromosomes but increases the number of metacentrics as compared to the number of acrocentrics.

Number of Chromosome Arms.   The basic number of chromosome arms, the "nombre fondamental" (N. F.), is remarkably constant for the various groups of amphibians and reptiles. The N. F. of any stock is the number of all the chromosome arms, including the arms of both microchromosomes and macrochromosomes. To calculate the N. F. one simply adds the number of acrocentrics to twice the number of metacentrics. Some related forms that are quite different in chromosome numbers have the same N. F. For example, the only two caecilians that have been examined cytologically, *Ichthyophis glutinosus* and *Uraeotyphus narayani,* have 42 and 36 chromosomes, respectively, but the former has 32 I-shaped acrocentrics and 10 V-shaped metacentrics, giving it an N. F. of 52, whereas the latter has 20 I-shaped acrocentrics and 16 V-shaped metacentrics so that it likewise has an N. F. of 52.

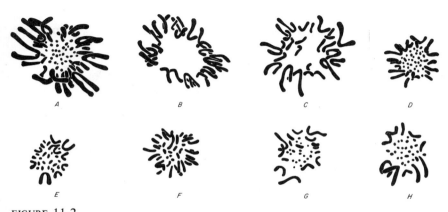

FIGURE 11-2
Diploid chromosomes of several genera of amphibians and reptiles:
(A) Cryptobranchus; (B) Triturus; (C) Rhacophorus; (D) Clemmys; (E) Alligator; (F) Gekko; (G) Elaphe; (H) Naja.

There has been little success in the correlation of shifts in karyotypes with individual evolutionary modifications of structure. Nevertheless, karyotypic studies strikingly confirm some of the conclusions derived from paleontological and anatomical studies. They support the primitive position assigned to the families Hynobiidae, Cryptobranchidae, Ascaphidae, Discoglossidae, and Pipidae. At a lower systematic level, they have been used, for example, to confirm the division of the lizard genus *Anolis* into two major species groups. They support the distinctness of the South American toads, *Bufo maximus* and *Bufo arenarium,* from their North American and European congeners. And they offer much hope for the future clarification of some perplexing problems of classification.

Genetics

Herptiles are relatively slow breeders. Many of them take two or more years to reach maturity. In addition, the offspring of the amphibians usually pass through an aquatic larval stage before they metamorphose to a point where their characters can be compared to those of their parents. Also, it is usually more difficult and expensive to raise large numbers of herptiles in a laboratory or under controlled conditions than it is to raise fruit flies or mice. For these reasons, detailed genetic studies of amphibians and reptiles have lagged, but a start has been made. Most of the work so far has been concerned with the inheritance of color pattern.

Genetic Studies on Amphibians. The little Greenhouse Frog of the West Indies *(Eleutherodactylus ricordi planirostris)*, which has become so widespread in Florida, has two distinct dorsal patterns: striped and mottled (see Fig. 11-3). When two striped individuals are mated, their offspring are either all striped or some are striped and some mottled; the same is true of cross-

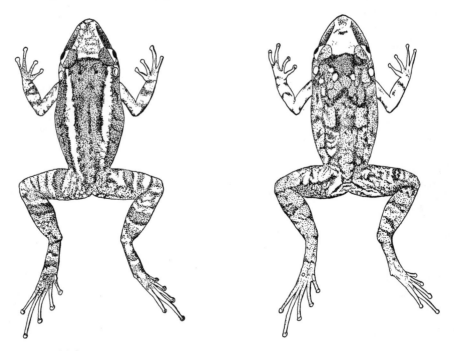

FIGURE 11-3
Striped (*left*) and mottled (*right*) patterns of the Greenhouse Frog, *Eleutherodactylus ricordi planirostris*. [From Goin, *Studies on Eleutherodactylus*, University of Florida Press, 1947.]

breeds of striped and mottled, but when two mottled frogs are mated, only mottled young are produced. Apparently, the gene producing the striped condition *(S)* is dominant and the gene for the mottled condition *(s)* is recessive. A random sample taken to determine the phenotypic ratio of the population at Gainesville, Florida, indicated that 0.2443 of the population were mottled and thus presumably homozygous recessives. Once the number of homozygous recessive animals in a population is known, it is possible, by means of the Hardy-Weinberg equilibrium formula, to determine the relative frequency of the two genes in the population. The equilibrium formula is:

$$q^2 = SS$$
$$2q(1-q) = Ss$$
$$(1-q)^2 = ss$$

Since we know that $ss = 0.2443$, we can solve the last equation to find the value of q and then substitute this value in the other equations. Thus it is possible to compute the probable gene distribution in the Gainesville population.

$$(1-q)^2 = 0.2443 \qquad ss = 24.43\% \qquad \text{Mottled}$$
$$2q(1-q) = 0.4999 \qquad Ss = 49.99\%$$
$$q^2 = 0.2557 \qquad SS = 25.57\% \qquad \text{Striped}$$

Also, if the frogs mate at random, striped and mottled patterns should appear in the offspring in about the same proportion as in the parent population. Of 1395 frogs hatched from eggs collected in the same area, 354 were mottled and 1041 were striped. Theoretically, of these 1395 offspring, 24.43 percent or 341 should have been mottled and 75.57 percent or 1054 striped. The close correspondence between the theoretical and actual patterns in the offspring is in agreement with the view that mating is random.

Similar studies on other species of *Eleutherodactylus* indicate that the striped pattern is also inherited as a dominant trait in such forms as *E. alticola, E. nubicola, E. gossei, E. pantoni,* and *E. nasutus.* It seems that this pattern, so widespread in the genus, is due simply to homologous genes in many of the species. Other patterns found in frogs of this genus, such as the presence of a narrow, middorsal stripe, or of a shield-shaped, light area on the back (called "picket") apparently are likewise expressions of simple color pattern genes.

The uniform dorsal coloration in the frog described as *Rana burnsi* is due simply to a single dominant gene that is allelomorphic to the recessive gene causing the spotted pattern in the Leopard Frog, *Rana pipiens.* Not only this, but there is strong evidence that the dorsal spotted patterns of two quite distinct species, *Rana areolata* and *Rana palustris,* are expressions of the same gene and that these three species have this corresponding gene locus in common. In *R. areolata, R. palustris,* and the spotted "wild type" of *R. pipiens,* the genotype is pure recessive *(bb)* but in the *R. burnsi* form of *R. pipiens* it is *Bb.* The form *R. burnsi* is rather rare. It is restricted geographically to northern Iowa, southern Minnesota, and northwestern Illinois, and constitutes less than 5 percent (often less than 1 percent) of the total *R. pipiens* populations in the areas in which it occurs. The absence of any *BB R. burnsi* frogs among those tested in breeding experiments apparently results from the rarity of the gene in the population. Parenthetically, it might be stated that the form *R. burnsi* obviously should not be considered a separate species.

Sometimes the inheritance of color pattern has a more complex basis. The inheritance of green color in the Pacific Tree Frog, *Hyla regilla,* is determined by two dominant genes, *O* and *G.* If either of these genes is present as a homozygous recessive, green does not develop. These frogs may also show a red dorsal stripe whose presence is determined by a single gene with nonred dominant to red.

Eight to twelve percent of individuals of the frog *Rana cyanophlytis* collected around Delhi, India, have two or more vertebrae fused. This character has been shown to be due to a single dominant gene. Embryos heterozygous for this gene are able to develop normally at high temperatures that are lethal to homozygous recessive embryos. This suggests a balanced polymorphism in which the heterozygote has an advantage over the homozygote.

Genetic Studies on Reptiles.   In reptiles as in amphibians genetic analysis has dealt largely with the inheritance of color pattern. The population of Garter Snakes *(Thamnophis s. sirtalis)* around Lake Erie includes a number of black individuals as well as ones with the normal striped pattern. In some places the black ones make up as much as a third of the population. It has been shown rather conclusively that the melanistic pattern is due to a recessive gene *"b,"* so that Garter Snakes of the genotypes *BB* or *Bb* have the normal striped pattern but those of the genotype *bb* are black.

A more striking instance among snakes, albeit admittedly less well understood, is that of the Kingsnakes of California *(Lampropeltis getulus californae)* in which a striped form breeds with a ringed form. Enough data have been accumulated to demonstrate that both striped and ringed offspring are produced by mothers of both types, thus indicating a mendelian basis for the patterns. The evidence is too scanty as yet to determine which of the two patterns is dominant and whether or not there is random mating in the population. Finally, some of the offspring have a somewhat aberrant pattern that is obviously allied to, although different from, the striped pattern. Whether this is due to a modifying gene is not known.

SPECIATION

The appearance and spread of mutations through a population is not in itself enough to cause speciation. Dr. Ernst Mayr has stated that "the problem of the origin of the species is a problem of the origin of discontinuities, for although evolutionary change alone may lead to modifications of perviously existing species it cannot lead to their multiplication." This is an excellent summary of the situation. If a discontinuity develops between two parts of a population and this discontinuity is maintained long enough to allow for genetic reconstruction and the development of isolating mechanisms, speciation has occurred.

The first step in the study of speciation is, therefore, the search for the causes of discontinuities. They may arise either:

1. Instantaneously, by means of polyploidy
2. Gradually, by:
    a. Geographic separation, in which the populations occupy different areas and are separated either by distance or by a physical barrier such as a mountain range.
    b. Ecologic separation in which the populations occupy essentially the same area but different habitats within the area.

Speciation by Polyploidy.   In the past few years, it has become obvious that speciation by polyploidy is more common among amphibians and reptiles than had been previously supposed. Triploid populations have been reported in salamanders of the genus *Ambystoma* and in the lizard genera *Cnemidophorus, Lacerta,* and *Leiolepis.* The triploid *Ambystoma* strains apparently arose through hybridization between the Jefferson Salamander, *Ambystoma jeffersonianum,* and the Blue-spotted Salamander, *Ambystoma laterale.* The triploids are all female and they reproduce gynogenetically. The eggs are formed by mitosis, and development is initiated by penetrance of sperm from either *A. jeffersonianum* or *A. laterale,* but the sperm pronucleus does not fuse with the egg pronucleus and contributes no genetic material to the developing embryo. The triploid species of *Cnemidophorus* reproduce parthenogenetically. Two South American frogs have been shown to be polyploid. *Odontophrynus americanus* is tetraploid and *Ceratophrys dorsata* is octoploid. These two species are bisexual. Recently it has been demonstrated that the Gray Tree Frog *(Hyla versicolor)* of eastern and central United States is also tetraploid.

Geographic Discontinuities.   A population may have a wide geographic range. Since environmental conditions differ from one part to another of the range, the members of the population are exposed to different selective forces in different areas. When the change in conditions is gradual, as the increase in temperature as one moves from north to south in the northern hemisphere, characters in the population may also show a gradual change. For example, the average number of ventral plates of the Eastern Kingsnake *(Lampropeltis g. getulus)* increases from north to south. In southeastern Virginia it is 211; in North Carolina, 212; in South Carolina, 213; in southeastern Georgia, 216; and in northern Florida, 218. Such a character gradient is called a "cline." The Kingsnakes show similar clines in other characters, so that the part of the population at one end of the range differs noticeably from that at the other end.

In many species, characters do not show the smooth curve of clinal change. They are relatively constant over a part of the range, then shift rather abruptly. This may be because the environmental factors also change abruptly, as in the transition from woodland to open grassland, or because some partial barrier reduces the flow of genes from one segment of the population to another. (Differences in environment may themselves act as partial barriers by limiting the amount of contact, and hence the degree of interbreeding between the parts of the population.) When the characters by which one segment of a population differs from another change abruptly rather than gradually, the two may be named as subspecies or geographic races.

Ecologic Discontinuities.   Geographic and ecologic discontinuities really differ only in degree, since ecologic discontinuities are microgeographic as well. Thus, on Jamaica, frogs of the genus *Eleutherodactylus* have certainly undergone ecologic speciation, but this speciation is at the same time associated with minor geographic differences. *Eleutherodactylus gossei* is widespread in the lowlands and on the dry lower slopes of the mountains. In the very humid cloud forest, which occurs at elevations of about 1200 meters, a larger, dark-bellied form, *E. nubicola,* is present. Above this, ranging into the wind-scrub on the top of Blue Mountain Peak, occurs *E. alticola,* smaller than *E. nubicola* and much less variable in color pattern than the other two. Obviously, the cloud forest is not only ecologically very different from the dry lower slopes but is also above them in a geographic sense. It seems useful, though, to maintain the term ecologic (sympatric) speciation to differentiate cases of this sort from the speciation resulting from discontinuities produced by simple geographic distance (allopatric speciation). Although all ecologic speciation is microgeographic, it is also true that geographic speciation is also ecologic since pure geographic distance seems invariably to involve some change in ecologic conditions.

Genetic Reconstruction

Once discontinuities have developed in a population, two things in addition to mutation can bring about genetic reconstruction of the isolated forms. One of these is natural selection. Since no two areas are ever exactly alike, the two parts of the population are exposed to different selective forces and hence come to differ from each other.

The second factor leading to genetic reconstruction is random fixation. The individuals of a population vary. Many genes are present in all members, but other genes are present in some individuals and not in others. Thus some of the *Eleutherodactylus* discussed above have the gene for the striped condition and some do not. The sum of the genes present in a population is known as the gene pool. When a segment of a population is separated from the rest, mere chance determines how its gene pool differs from that of the parent population in relative proportions of genes. A gene rare in the one may be relatively common in the other, or it may be absent entirely. If one or both of the populations is small, the gene pools will come to differ further from each other through the process known as genetic drift. If only a few members of the population carry a given gene, and if they, by chance, fail to reproduce successfully, the gene will be eliminated from the population. If the population is very small, even relatively common genes may be eliminated. Conversely, a gene that is common in the population may come to be fixed, that is, to be present in all members of the population. In very large populations, in the absence of selection, the relative proportions of genes in the gene pool remain about constant.

There have been many excellent studies of geographic variation in the herptiles. We can determine from them some of the characters by which the populations differ, but too often we can do no more than guess as to whether these characters were fixed by selection or by random genetic drift within the population. Studies of the genetic differences between related species, and of the selective values of the genotypes under different environmental conditions, remain an acute need.

Isolating Mechanisms

Differences of many kinds may develop between isolated populations, but only those differences that bring about genetic isolation lead to speciation. These we call isolating mechanisms. Once such mechanisms have developed, the formerly isolated populations may come to occupy the same area through a breakdown of barriers, but they will remain differentiated. Isolating mechanisms are of various kinds. Some of them are outlined below.

    I. Factors that inhibit crossing
       A. Breeding behavior differences
         1. In breeding sites
         2. In time of breeding
         3. In courtship patterns
       B. Physical differences
         1. In recognition characters
         2. In form of genitalia and related structures

    II. Development of genetic incompatibility
       A. Hybrid sterility
       B. Hybrid inviability
       C. Primary sterility

Different Breeding Sites.　Good examples of differences in choice of breeding sites have as their *sine qua non* detailed life history studies, of which, unfortunately, there have been all too few. Still, species pairs are known in which the adults live together during most of the year, but move to different sites in the breeding season. Of the two toads found in England, the Common Toad *(Bufo bufo)* breeds in moderately deep water in ponds and canals, whereas the Natterjack *(Bufo calamita)* breeds in shallow water, often in small puddles only an inch or two deep.

Different Breeding Seasons.　At Bloomington, Indiana, the American Toad *(Bufo americanus)* begins to breed during the last week in March, but Fowler's Toad *(Bufo woodhousei fowleri)* does not start until the middle or end of April. The two forms also differ in their calls and, to a large extent, in their choice of breeding sites. There is some overlap of the breeding seasons and during this brief period the two occasionally interbreed. Furthermore, of the toads that breed in this period of overlap, a large percentage of the individuals are more or less intermediate morphologically between the two species and their calls are apt to be somewhat intermediate. None of the crosses discovered during the period of overlap involved individuals near the mean of variation of either of the respective species. Isolation between the two forms is thus not complete. It has been suggested that they may have come together in relatively recent times, perhaps because ecologic conditions in the region have been altered by human activities. Unless there is some selective pressure against the hybrids, the two species may eventually merge.

    It may well be that two of the hognose snakes, *Heterodon platyrhinos* and *Heterodon simus,* are separated in part by a mechanism of this sort. *H. platyrhinos* emerges fairly early in the spring whereas *H. simus* is definitely a hot weather snake. Perhaps *H. platyrhinos* copulates earlier in the year than *H. simus,* but the actual time when *H. simus* copulates is still unknown. Studies of this sort are also needed.

Different Courtship Patterns.   Male turtles of the species *Chrysemys scripta* go through a remarkably different Liebespiel from that of the males of *C. floridana.* In the former the male swims backward facing the female, stroking her face and chucking her under the chin with the long nails of his front feet. In the latter species both face in the same direction and the male swims just above the female, his head bent close to hers. He then turns his tail down and under the rear margin of her shell; at this point, the "Liebespiel" ends and the two sink slowly to the bottom, where presumably copulation takes place.

Different Recognition Characters.   One has only to listen to a few frog choruses to realize that related species may breed in the same pond but have quite different voices. A number of studies have shown that female frogs are able to discriminate between the calls and move preferentially to the males of their own species. Two species of Gray Treefrogs *(Hyla versicolor* and *Hyla chrysoscelis)* are indistinguishable in appearance and may breed in the same ponds. The call is a loud, resonant trill but is much lower in pitch in *H. versicolor* than in *H. chrysoscelis.* Tests with tape recordings have shown that a female is able to distinguish between the two and, when given a choice, will swim toward the sound of the voice of her own species. There is a high degree of genetic incompatibility between the two species, in spite of their great similarity. It seems probable that *H. chrysoscelis* is the diploid form from which the tetraploid *H. versicolor* arose.

That size likewise may be effective as a barrier to crossbreeding is indicated in part by the geographic distribution of many forms. The salamander genus *Plethodon* of the eastern United States includes many species pairs that live and breed in the same woods but are different in size. Since they do not depend on voice for recognition, and since they are usually found under rocks, logs, or leaf litter where sight would be of little use, it may well be that the difference in size is an important isolating mechanism. Detailed life history studies will probably some day give us a long list of isolating mechanisms based on recognition characters.

Different Genital Structures.   Two Asiatic species of lancehead snakes of the genus *Trimeresurus* are strikingly similar both in color pattern and in external morphological features. The hemipenis of *T. stejnegeri* is short, thick, not deeply forked, and bears heavy spines. The hemipenis of *T. albolabris* is longer, more slender, more deeply lobed, and is without spines. The cloaca of a female *T. stejnegeri* is shorter, not so deeply lobed, and has

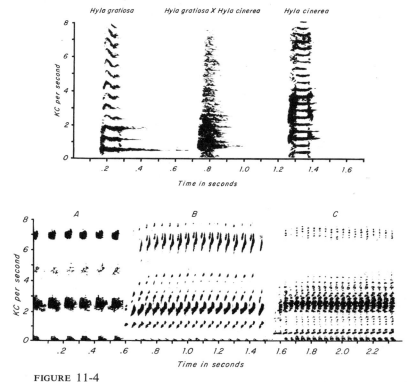

FIGURE 11-4

Sound spectrograms of the voices of various tree frogs of the genus *Hyla*.
It is probable that differences in voice act as isolating mechanisms in frogs.
Above, left to right: the mating calls of *H. gratiosa,* a hybrid *H. gratiosa*
× *H. cinerea*, and *H. cinerea*; all were recorded near Archbold Biological
Station, Highlands County, Florida. *Below*, the mating calls of three hylids:
(*A*) *H. femoralis*, a portion of call recorded at Orange Springs, Florida;
(*B*) *H. versicolor*, entire call recorded at Hazelhurst, Wisconsin; (*C*) *H.
arenicolor*, entire call recorded at Cave Creek, Arizona. [Spectrograms
from Bogert, in *Animal Sounds and Communications*, A. I. B. S., 1961.]

much thicker walls than that of *T. albolabris*. It seems likely that the long
spines on the hemipenis of *T. stejnegeri* might damage the thin-walled cloaca
of *T. albolabris* during copulation, while the slenderness and lack of spines
of the hemipenis of *T. albolabris* might prevent its being held in place in the
cloaca of *T. stejnegeri*. It is not known whether these two species attempt to
crossbreed in nature, but it may be that here the differences in genital structure are the primary isolating mechanism.

Genetic Incompatibility

The accumulation of genetic differences in discontinuous populations may
lead eventually to the development of genetic incompatibility between the

two forms. There are many gradations of genetic incompatibility: individuals of the two populations may be able to crossbreed to produce hybrid offspring that survive to maturity but are unable to reproduce successfully; the hybrids may show a lessened degree of viability so that few or none survive to maturity; the zygotes may fail to develop beyond the early stages; or there may be no embryonic development at all.

The widely distributed Leopard Frog complex (*Rana pipiens* and related forms) ranges from Canada to Central America and from the Atlantic to the Rocky Mountains. In different parts of the range, natural selection has brought about genetic reconstruction to modify the physiological adaptations of the embryos. These embryos are narrowly adapted to the environmental conditions under which they normally develop. They differ in temperature tolerance, rate of development, and temperature coefficient of development. When individuals from adjoining parts of the range are crossed, the embryos develop in a perfectly normal manner, but the greater the geographic distince, the greater the percentage of abnormalities in the embryos. Vermont males crossed with New Jersey females produce normal embryos, whereas embryos of the reciprocal cross are either normal or very slightly abnormal. When Vermont females are crossed with males from the closely related species, *R. sphenocephala,* from Englewood, Florida, there is a marked retardation in the rate of development and the head of the embryo is greatly enlarged. The reciprocal cross produces embryos that show marked retardation in development and extreme reduction in the size of the head. Also, there is a high percentage of mortality and hence a high degree of hybrid inviability between these two populations. When even more distant members of the species group are crossed, the percentages of abnormal and inviable offspring increase and cellular disintegration is apt to begin early in embryonic development. Crosses between *R. pipiens* from Wisconsin and *R. berlandieri* from Tamaulipas, Mexico, fail to produce any offspring that survive to the time of metamorphosis.

Crosses between species belonging to the Wood Frog group have produced results ranging from the production of hybrids that reach maturity to complete sterility. This group comprises a chain of species and subspecies extending from western Europe across northern Asia, through Alaska and Canada, to the eastern coast of North America. Crosses of two of the European species *(Rana temporaria* and *R. arvalis)* produce hybrids that survive to maturity, as do crosses of two Japanese forms *(R. japonica* and *R. temporaria ornativentris)*. In crosses between *R. temporaria* males from western Europe and *R. sylvatica* females from eastern North America, development stops at the late blastula stage, and in the reciprocal cross there is no embryonic development at all. Apparently, genetic incompatibility between the two end species of the chain has reached the stage of causing complete sterility.

On the other hand, complete hybrid viability may exist between quite distinct, sympatric species. In eastern North America, *Rana pipiens* and *R. palustris* (the Pickerel Frog) produce hybrids that can be carried to transformation, although these species do not normally attempt to mate in nature. This is also true of the sympatric *Rana temporaria* and *R. arvalis* mentioned above. These species maintain their distinctness, not through genetic incompatibility, but through other isolating mechanisms.

When genetic incompatibility has developed between two forms and they come to occupy the same or contiguous areas, the individuals of one form that do not attempt to breed with individuals of the other will be the ones whose offsprings will survive. Therefore natural selection may be expected to bring about a parallel development of isolating mechanisms that inhibit attempts to cross.

In conclusion, two points about isolating mechanisms should be emphasized: (1) no single mechanism ever works alone; (2) mechanisms develop through genetic reconstruction of populations that are initially separated ecologically or geographically.

READINGS AND REFERENCES

Blair, W. F., ed. *Vertebrate Speciation.* Austin, Texas: University of Texas Press, 1961.

Dobzhansky, T. *Genetics and the Origin of Species.* New York: Columbia University Press, 1937. (Emphasizes the genetic basis of speciation. All of the books on this list are general works, but most of the principles discussed in them apply to the herptiles.)

Huxley, J. *Evolution: The Modern Synthesis.* New York: Harper Brothers, 1942. (A very comprehensive synthesis of the many, complex, and interrelated factors involved in speciation.)

Matthey, R. *Les Chromosomes des Vertebres.* Lausanne: F. Rouge, 1949. (Contains much information about the karyotypes of the herptiles.)

Mayr, E. *Systematics and the Origin of Species.* New York: Columbia University Press, 1942. (A companion volume to that of Dobzhansky, discussing both the contributions of systematic studies to an understanding of evolutionary processes, and the elucidation of systematic problems by modern theories of the mechanisms of speciation.)

———. *Animal Species and Evolution.* Cambridge, Mass.: Harvard University Press, 1963.

White, M. J. D. *Animal Cytology and Evolution.* Cambridge: Cambridge University Press, 1945. (Deals with the structure and evolution of chromosomes and their bearing on evolutionary problems.)

12

GEOGRAPHIC DISTRIBUTION

Since man first began to explore the world around him, he has been aware that different kinds of animals live in different regions: crocodiles are found in the River Nile but not in the Thames; many snakes live in Europe but there are none in Ireland. As long as the doctrine of Special Creation held sway, it was easy to explain these facts by saying that each region had its own fauna, created especially for it and adapted to it. But this explanation was never really satisfactory. For one thing, there is too much overlap: not all animals are restricted to a single region. Many of the animals the early explorers found in the New World were different from anything they had ever seen, but others were strikingly similar to those they had known at home. Then it became obvious that animals can adapt well to regions in which they do not naturally occur. The horses the Spanish Conquistadors brought to America throve in the new environment. Florida is obviously a fine place for the Greenhouse Frog *(Eleutherodactylus ricordi planirostris),* which was first reported in Key West in 1863 and 80 years later had spread almost to the northern border of the state. When the true nature of fossils was finally recognized, and their orderly succession became apparent, men realized that some animals once lived in regions in which they are no longer found and that many forms have become extinct. Dinosaurs once roamed every continent, but they no longer exist. It became necessary to postulate a whole series of special creations, with the next to the last one presumably destroyed by the Biblical Flood. But this attempt to maintain the doctrine of Special Creation eventually became absurd.

It was Darwin's observations on the distribution of animals in South America and the nearby Galápagos Islands that first turned his thoughts to the idea of evolution. Now the two studies suppplement and reinforce one another: the pattern of distribution of animals is one of the strongest arguments in support of evolution and offers many clues to the course evolution has taken; and, conversely, to understand the present distribution of animals we must take into account not only their ecological requirements but also their evolutionary history. The correlation of the geographic distribution and evolutionary history of some groups that are well represented in the fossil record may enable us to form sound ideas concerning the evolution of other groups by an inspection of their geographic distribution.

FACTORS IN GEOGRAPHIC DISTRIBUTION

The geographic distribution of any group of animals is the result of the interplay of two sets of factors, extrinsic and intrinsic:

    I. Extrinsic factors
       A. Distribution of favorable environments
       B. Changes in environments through geologic time
          1. Climatic
          2. Biotic
       C. Formation of highways permitting dispersal or of barriers to dispersal
   II. Intrinsic factors
       A. Physiological requirements of group
       B. Time and place of origin
       C. Potential rate of spread
          1. Biotic potential
          2. Vagility
       D. Genetic plasticity of group

Extrinsic Factors

It is obvious that, for a group of animals to occur in a region, environments must be available that are suited to the needs of that particular group. We do not expect to find the moisture-loving salamanders in a desert. *Salamandra salamandra* has been able to reach northern Africa, but it has not been able to extend its range into the Sahara. The cold arctic regions are not suited to the ectothermic herptiles and few of them are found there. The True Frogs *(Rana)* reach the Arctic in both the Old and New Worlds and a salamander *(Hynobius keyserlingi)*, a lizard *(Lacerta vivipara)*, and a snake *(Vipera berus)* extend that far north in the Old World, and that is all. A favorable environment implies not only suitable physical features but also favorable biotic conditions. There must be a nice balance between predator and prey species, hosts and parasites, competitors and food.

Most environments are stable, however, only over relatively short periods of time in the geologic sense. One of the best known of the changes that have taken place is the warming of the northern hemisphere in the last ten thousand years. This is evidenced by the retreat of the glaciers since the close of the Ice Age. The process is apparently continuing to the present day and many groups of animals are still extending their ranges northward. Further back in geologic time, the coal measures in Pennsylvania show that semi-tropical swamps once flourished in what is now a hilly, well-drained, temperate region.

Climatic factors of the environment are not the only ones that change, biotic factors are also constantly shifting since the forces of evolution are continually at work on all forms of life. New sources of food become available to animals that are able to take advantage of them, new enemies appear which must be evaded. Perhaps most important of all, new and better adapted competitors for food and breeding sites either move into the area or evolve within it.

Although the major continental land masses have probably remained relatively constant, at least since the appearance of the modern groups of amphibians and reptiles, there have been many changes in the connections between them. South America was cut off from North America by an arm of the sea for the greater part of the Cenozoic Era. During this time distinct faunas evolved in the two regions. When the land connection was reestablished during the Pliocene, it became possible for North American forms to invade South America and vice versa. The mingling has been very incomplete and the two faunas are still essentially different. Thus only one group of salamanders (*Bolitoglossa* and its allies of the family Plethodontidae) has invaded South America. In contrast, the North American continent has apparently been connected in the past with Asia across the Bering Straits at times when the climate was sufficiently mild so that this land bridge proved a suitable highway for the dispersal of many groups of amphibians and reptiles. The fauna of temperate North America resembles that of Asia more closely than it does that of South America.

Obviously, a given topographic feature may serve as a highway of dispersal for some forms and as a barrier to others, depending on the physiological requirements of the groups involved. Broad lowland valleys are barriers to salamanders that are adapted to dwelling on mountain tops but many serve as highways to other species such as the toads.

Intrinsic Factors

Even closely related forms often show differences in their physiological requirements. Furthermore, some animals have broad ecological tolerance and are able to adapt themselves to conditions over a wide area while others are very limited in ecological tolerance and hence are restricted to a narrow range. This sort of difference is probably reflected in the distribution of two of the North American Rat Snakes: the ecologically adaptable *Elaphe o. obsoleta* ranges from Ontario and northern New England south to Georgia and west to Minnesota and Texas whereas *E. subocularis* is limited in range to the arid region of Trans-Pecos Texas, southern New Mexico, and adjacent Coahuila.

Besides the animal's physiological requirements, which determine which environments it may occupy and what highways of dispersal are available to it, the time and place of origin of a group play an important part in determining its geographic distribution. If a group arises in a region that is blocked from an adjacent region by some barrier, it will not be able to spread into that region even though there may be environments there that are well suited to its needs. The frog family Leptodactylidae is very widespread in South America and contains many genera, but only three of these are present in the United States; this is probably because it was only recently, in the geologic sense, that a passageway was opened.

If a group is of very recent origin, it may not have had time to spread very far from its center of origin. How fast a group will spread depends in part upon its biotic potential. Animals that are capable of producing large numbers of offspring in a relatively short time will, other things being equal, be able to occupy new areas more rapidly than forms with a low rate of reproduction. The vagility (inherent power of movement) of the species may also have its effect on the rate of speed. Both of these factors are probably relatively minor since many forms with low biotic potentials and limited vagility have been able to occupy large areas of the earth's surface.

Finally, the genetic plasticity of the group will determine whether it will be able to occupy new environments, whether it can adapt itself to changes in the environment *in situ*, or whether it must either follow receding belts of its old environment or become extinct.

PATTERNS OF DISTRIBUTION

During their evolutionary history, animal groups pass through various stages. When a group first arises, it is few in number and occupies a limited area. It then goes through an expanding phase in which it spreads to occupy whatever territory is available to it. This is usually followed by a contracting phase in which the population is reduced and eventually extirpated over much of its former range. Since different groups have arisen at different times during the past and since, furthermore, they differ in the rates at which they pass through these stages, we must expect them to show differences in distributional pattern.

Expanding Populations

It is axiomatic that a population of animals must be adapted to the environment in which it is found or perish. But as time passes this population, or at least a part of it, is exposed to the harsh influences of new environments to which it is not adapted. This may come about in two ways.

Most animals are capable of producing many more offspring than could be expected to survive within the limits of the area occupied by the parent population. This means that overcrowding is an ever-recurring phenomenon in any but a decadent population. The animals near the periphery of the range are constantly being pushed into new and unfavorable environments. Undoubtedly, most of them perish. Stocks that have enough genetic plasticity are sometimes able to make the change so that their descendants become adapted to the new conditions and hence develop into incipient discrete populations. Under this process we have the primitive types occurring toward the center of the geographic range and specialized types developing at the periphery. This is the so-called rimfire pattern of distribution.

The frog genus *Eleutherodactylus* is widespread in northern South America and the species in this area seem to be relatively stable. At the periphery of the range, in the West Indies, the group is undergoing rapid speciation. Thus, on the relatively small island of Jamaica, besides two recently introduced forms, there are fourteen endemic species of *Eleutherodactylus*. (Species endemic to a region are ones that originated in and still occupy that region.) These fourteen species apparently evolved on Jamaica from three, or at most four, stocks that reached the island from the mainland.

Since the world's environments are constantly changing, albeit at times very slowly, this pattern of distribution must most commonly occur in relatively small areas and over relatively short periods of time.

The second way in which a population of animals may be exposed to a new environment is through changes in the climatic and biotic conditions within the area in which it lives. This results in a distribution pattern quite different from the one just discussed. Animals near the periphery of the range may be able to follow receding belts where the old conditions are extant. They may thus remain unchanged, whereas the forms toward the center of origin must become adapted to the new conditions or perish. Since most environmental changes over large areas take place rather slowly this is often possible. Here we have specialization taking place near the center of origin of the group, with the primitive types located at the periphery—the centerfire pattern of distribution. This process applies principally to larger, more inclusive groups occupying extensive areas over long periods of time.

The widespread frog family Ranidae shows this type of distribution. Africa seems to have been a center of differentiation for the group. Six of the eight subfamilies are found there and four are found nowhere else. The rather unspecialized genus *Rana* reaches Europe, North and South America, Asia, and Australia.

Expanding populations generally exhibit continuity of range. When this is so, the lines of continuous distribution should lead toward the center of origin of the group as a whole. Thus, if the range of an animal group (species, genus, family) is continuous and we find the specialized types at the periphery (in young, small groups) or concentrated near the center (in older, larger groups), we have good indications that the center of origin of the group is somewhere near the geographic center (that is, the center of available migration routes) of the group.

Relict Populations

Since the history of the fauna of the earth has been one of extinction as well as of evolution, we must expect to find in addition to expanding populations many examples of decadent and relict stocks. Although the causes and rates of extinction are not completely understood, apparently subservience to later, better-adapted types plays at least as important a role as rigorous changes in the physical environment. Thus a population that originated and spread from a center of origin may well be eliminated by a later, more successful population originating at the same center. Hence, centers of origin may also become centers of extinction.

Except for certain island faunas, perhaps the best clue to a decadent population is discontinuity of range. Where there are gaps in the distribution, it means either that a stock has been introduced into a distant area or that it spread there under its own power and the links connecting it to the parent stock have been destroyed. The alligators, with one form in the southeastern United States and one in southeastern Asia, have a typical relict distribution. Apparently the alligators living today are remnants of an earlier, widespread stock; their fossil record confirms this.

Thus, where we find a group divided into isolated stocks, there is a strong probability that we are dealing with a receding rather than an expanding population and that these isolated stocks are relicts. Since the center of extinction of a group may be anywhere within its expanded range, it follows that in these receding populations the geographic center of the present distribution need not be near the center of origin of the group. In fact, a receding population may well be extinct at the place of origin. Here we must depend upon the fossil record to determine the center of origin.

Waif Populations

A waif population is one whose ancestors reached an area (usually an island) as stray or castaway individuals rather than through the normal dispersal of an expanding population. In times of flood, uprooted trees or segments of the river bank may be swept downstream and carried out to sea. They may be caught by ocean currents and eventually stranded on the shores of a distant island. Such a natural raft may harbor a clutch of eggs, a pregnant female, or a pair of individuals of the same species. If they survive the hazards of the journey, they may be able to establish a population on the island.

It is not always easy to tell whether a population is waif or whether it reached the island at a time when the island was still connected to the mainland. In either case, speciation may have occurred so that the population differs from any now found on the mainland. Islands that have been formed entirely by volcanic action or that have been completely submerged since their last previous connection with another land mass must *ipso facto* be populated by waifs.

In the past few thousand years, man has, either accidentally or purposely, carried many animals from one part of the globe to another. It is customary, though, to speak of populations arising from animals transported by human agencies as introduced rather than waif species.

The island of Bermuda is inhabited by a species of lizard, *Eumeces longirostris*, that differs greatly from all other species of the genus and is found in no other place; it is undoubtedly waif. The other herptiles of the island seemingly have all been introduced by man.

ZOOGEOGRAPHIC REALMS

With the development of a dynamic approach to zoogeography, several attempts have been made to divide the earth's surface into zoogeographic realms, broad areas characterized by general resemblances of the fauna within each realm and general differences between the faunas of one realm and another. One of the first of these attempts was by an ornithologist, P. O. Sclater, who, in 1858, recognized six major realms. His divisions were so carefully selected that they still form the basis of our divisions today. His terminology, however, has been somewhat modified. A recent and acceptable nomenclature for these regions follows (see also Fig. 12-1).

Arctogaea:
> Holarctic Region:
>> Palearctic subregion
>> Nearctic subregion
> Ethiopian Region:
>> African subregion
>> Madagascar subregion
> Oriental or Indian Region

Notogaea:
> Australian Region
> New Zealand Region
> Oceanic Islands Region

Neogaea:
> Neotropical Region

The Holarctic Region includes all of Europe and its islands, the northern belt of Africa, Asia north of the Himalayas, and all of North America north of the West Indies and an ill-defined zone in Mexico. It is readily and logically separable into the Palearctic subregion, the Old World portion of the Holarctic, and the Nearctic subregion, the New World portion of the Holarctic.

Madagascar, the Arabian peninsula, and all of Africa except the belt north of the Sahara form the Ethiopian Region. It is divided into a Madagascar subregion for that island and an African subregion.

The Oriental Region leads into the Australian Region by way of an archipelago. The islands of this archipelago form a series of stepping stones, so to speak, between the two. It is a matter of dispute between which two islands the boundary line should be drawn, if indeed the boundary is definable as a single line. The Oriental Region, as here defined, includes southern Asia south of the Himalayas and extends to the southeast through Ceylon, the Philippines, Sumatra, Borneo, and Java.

The Australian Region then comprises the islands south and east of those listed immediately above to and including Australia and the islands on its continental shelf.

North and South Islands of New Zealand and the smaller islands in the vicinity comprise the New Zealand Region.

The scattered islands of the Pacific are recognized as the Oceanic Islands Region.

Finally, the West Indies and Central and South America make up the Neotropical Region.

Although it is best not to attempt to define these regions too sharply, they are basically useful and the names of many of them are so firmly entrenched in the literature that we must have an understanding of them.

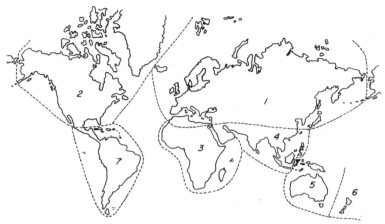

FIGURE 12-1

Zoogeographic regions and subregions: (*1*) Palearctic subregion:
(*2*) Nearctic subregion; (*3*) Ethiopian Region; (*4*) Oriental or Indian
Region; (*5*) Australian Region; (*6*) New Zealand Region; (*7*) Neotropical
Region. [Modified after Beaufort, 1951.]

Much difficulty has been caused by attempts to apply the names of the
regions to their faunas. Since none of the regions has a sharply limited,
uniform fauna, such attempts are at the outset bound to cause confusion.
For example, it has been shown that the herpetological fauna of the Nearctic
Region consists of a modern, circumpolar element, an "Old Northern"
element characteristic of the southern United States, and a South American
element representing a recent invasion from the south. Obviously we can
speak of the region, but not the fauna, as Nearctic.

ORIGIN OF THE FAUNA

There have been several attempts to select one area or another as the
major center of origin and dispersal for all present-day forms. There seems
to be no inherent reason, however, why all or most of the modern families
or other major groups should have originated in a single region. Just as
they arose at different times in the past, so did they probably arise in different places. The majority of them may have appeared first on the great
Holarctic land mass, simply because it is the largest of the major land areas.
The primitive and relict stocks in the southern hemisphere are thought to
have been derived from populations formerly occupying the nearly continuous land masses to the north, from which they have been eliminated by
the later, more advanced types and modern forms; thus remnant populations
persist toward the ends of the continental land masses that extend south-

ward like great peninsulas (see Fig. 12-2). Many reptile stocks exhibit this pattern of distribution. The side-necked turtles, for example, now live only in South America, Africa, Madagascar, Australia, and New Guinea, although they are known from the late Mesozoic of Europe, North America, and South America.

FIGURE 12-2
Diagram of the major land masses of the world, showing how the continental masses extend southward from the north polar region like great peninsulas.

Nonetheless, it is not likely that all of our modern herptiles are holarctic in origin. There have unquestionably been many secondary centers of evolution and some of them have undoubtedly been large enough and in existence long enough to have permitted the evolution of new major groups of animals. In fact evidence has accumulated to indicate that the Salientia originated in the Antarctic region and spread northward by means of the southward extending peninsulas of South America, Africa, and Australia. This may be why we now find ranids concentrated in Africa, bufonids in Africa and Asia, pipids in Africa and South America, and hylids and leptodactylids in South America and Australia. Indeed, with the exception of the relict ascaphids and discoglossids and the primitive pelobatids, only three unspecialized genera, representing three different families, have become widespread in the North Temperate Zone. These are *Rana* of the ranids, *Bufo* of the bufonids, and *Hyla* (including *Pseudacris*) of the hylids. It should be pointed out too, that whereas the side-necked turtles mentioned above have Mesozoic representatives in Europe and North America the earliest salientians are known from South Africa, Madagascar, and Patagonia.

GEOGRAPHIC DISTRIBUTION OF
LIVING AMPHIBIANS AND REPTILES

Despite the relative scarcity of their fossils, amphibians and reptiles are, in some respects, among the better animals to use in studies of geographic distribution. Like the vast majority of the world's animals they are ecto-thermic. The principles of distribution that they exemplify are thus apt to be of wider applicability than those shown by the better known endothermic birds and mammals. They are more limited in vagility than the winged insects and birds, or the widely ranging larger mammals. The outlines of their primary patterns of distribution are thus less apt to be blurred by rapid, wide, and essentially random dispersal. Because of their unshelled eggs and unprotected skins, amphibians are ecologically bound to regions where fresh water is available. They are quite intolerant of sea water. We can be reasonably sure then that they have not fortuitously crossed exten-sive areas of the sea or arid desert lands *en masse*. Where we have essentially similar faunas on either side of such a barrier we can safely assume that they reached those regions before that barrier formed.

Gymnophiona

No fossil caecilians are known, but the widely discontinuous distribution of present-day forms indicates that they are relicts of a once widespread stock. The group is essentially tropicopolitan and occupies four land masses: tropical America south to Buenos Aires; tropical Africa; the Seychelles Islands in the Indian Ocean; and the Oriental Region from Ceylon and India to Java, Borneo, and the southern Philippines. Caecilians are absent from the Lesser Sundas, Celebes, and the Australian region of the Orient, from Madagascar, and from all of the West Indies except Trinidad. Of the approximately thirty-four genera known, none is found on more than one of the major land areas occupied by the group.

Trachystomata

Modern trachystomes are found only in southeastern and central United States and northeastern Mexico, but fossil representatives of the order are known from the Cretaceous of Wyoming. The group was thus once much more widely spread in North America and is obviously relict today.

Caudata

The most striking characteristic of salamander distribution is the discon-tinuity of the major groups, reflecting wholesale extinction and the relict nature of the group as a whole. Of the seven recent families, all but two show more or less discontinuous distribution. In the family Cryptobranchidae, one genus *(Andrias)* is found in China and Japan and the other *(Cryptobranchus)* in the eastern United States. Although the caecilians are tropical, the salamanders are a north temperate group and have invaded the tropics only in Central America and northern South America; this invasion has been made by only one group *(Bolitoglossa* and allied forms of the Plethodontidae). At present the order is centered in North America. Six of the seven families are found there, with only the Hynobiidae being absent. Two families, Ambystomatidae and Amphiumidae, are restricted to North America.

Anura

The flourishing group of modern amphibians is that of the frogs and toads. They are abundant on all continents; the Oriental Region, Africa, and South America are very rich in them. Two of the most primitive families show the typical disjunct distribution of relict forms; the Ascaphidae are found in northwestern North America and New Zealand, and the Discoglossidae in temperate Europe and Asia and the Philippines. It is interesting that these primitive frogs are not found in Australia although the presence of an ascaphid on New Zealand suggests that this family once occurred there.

Today's large, vigorous families—Ranidae, Bufonidae, Hylidae, and Leptodactylidae—are all basically southern in distribution as are many of the smaller families—Pipidae, Phrynomeridae, Atelopodidae, and Centrolenidae. The most widespread family is the Ranidae which has at least one representative on all of the major land masses except Greenland and New Zealand.

Testudinata

The turtles are a rather ancient and conservative group; several of the present families show discontinuities in distribution. Our knowledge of how they got to their current places of abode is better than for many of the other groups of herptiles, simply because they leave a better fossil record. Thus the Carettochelyidae are at present found only in New Guinea but their fossils are known from Europe, Asia, and North America. Evidently the recent form *(Carettochelys insculpta)* represents a much reduced relict of a once widespread stock. The primitive side-necked turtles show the typical disjunct distribution of relict forms: the Pelomedusidae are found in Africa, Madagascar, and South America; the Chelidae in Australia, New Guinea, and South America. The most flourishing and widespread family is the Testudinidae which includes the pond turtles and land tortoises; it is

almost cosmopolitan in distribution, only Australia being without representatives.

Rhynchocephalia

There is only one living form of this ancient order. *Sphenodon* is now known
only from a few islands off the coast of New Zealand but the family goes
back to the Lower Triassic and is known in fossil form from Africa, Europe,
Asia, and North America.

Lacertilia

Although lizards are particularly abundant in the tropics, they occur on all
the continents, and one, *Lacerta vivipara*, reaches well above the Arctic
Circle in Scandinavia. One striking thing about the distribution of the families of lizards is that there apparently has been less exchange between the
faunas of the Old World and the New World than has taken place in the
other major groups of herptiles. Of nineteen families, only five are found
in both, nine are restricted to the Old World, and five to the New. Interest-
ingly enough, the most primitive infraorder, the Gekkota, is also one of
the most widely distributed. Members of it are found in tropical regions
throughout the world. Apparently these lizards are peculiarly liable to for-
tuitous introduction; the original distribution of some of them may never
be ascertained. Some lizard families that have restricted distributions at
present are apparently relicts. Thus the monitor lizards of the family Varanidae, which are now found in the tropics and some warm temperate regions
of Africa, Asia, and Australia, are known as fossils from North America
and Europe. On the other hand, some of the families of lizards that have
limited distribution are probably quite recent and have not yet been able
to spread far from their centers of origin.

Serpentes

The geographic distribution of the snakes is, in a sense, the most difficult
to analyze. This is because the family Colubridae, which contains about
two-thirds of the world's living snakes, is really not well known systemat-
ically. Although the genera and species have been defined fairly adequately,
the relationships of the genera within the family have not been determined
satisfactorily.

The snakes, like the lizards, are widely distributed and occur in habitable
places on all continents, but are absent from New Zealand and Ireland.
They are most abundant in the tropical regions of the Old World. All but
one of the eleven families currently recognized, and most of the subfamilies,

are found in the Oriental Region and three families are restricted to it. Eight families occur in the New World but only one is found there exclusively.

Amphisbaenia

The ringed lizards are found mainly in South America, the West Indies, and Africa. One form, the Worm Lizard *(Rhineura floridana)* occurs in Florida, and the genus *Bipes*, with three species, is present in western Mexico. One species of ringed lizard is found in Spain and several are found in Turkey and Asia Minor. The fossil record indicates that ringed lizards were once more widespread in the northern hemisphere than they are today.

Crocodilia

The Crocodilia are mainly tropicopolitan in distribution, but one genus, *Alligator*, inhabits the north temperate zone, occurring in the southeastern United States and southeastern Asia.

READINGS AND REFERENCES

Beaufort, L. F. de. *Zoogeography of the Land and Inland Waters.* New York: Macmillan, 1951. (A sound, descriptive, regional zoogeography.)

Darlington, P. J., Jr. *Zoogeography: The Geographical Distribution of Animals.* New York: John Wiley and Sons, 1957. (The modern definitive work on zoogeography.)

Dunn, E. R. "The Herpetological Fauna of the Americas." *Copeia,* vol. 1931, no. 3, 1931. (A classic paper on reptile distribution in the New World.)

Matthew, W. D. "Climate and Evolution." *Annals of the New York Academy of Sciences,* vol. 24, 1915. (Reprinted as a *New York Academy of Sciences Special Publication,* vol. 1, 1939. Modern zoogeography begins with Matthew's great work. It should be familiar to all who are interested in the distribution of animals.)

Reig, O. A. "Nuevos Datos y Nuevas Hipótesis Sobre la Cenogénesis de los Tetrapodos Sudamericanos." *Physis,* vol. 23, no. 65, 1962.

Simpson, G. G. "Turtles and the Origin of the Fauna of Latin America." *American Journal of Science,* vol. 241, 1943. (A stimulating zoogeography paper by a paleozoologist.)

Smith, M. A. *The Fauna of British India.* Vols. I-III. London: Taylor and Francis, 1931-1943.

Wallace, A. R. *The Geographical Distribution of Animals.* London: Macmillan, 1876. (The original definitive work on zoogeography.)

13

CAECILIANS, TRACHYSTOMES, AND SALAMANDERS

Caecilians, sirens, and salamanders are treated together in this chapter simply as a matter of convenience. They are mostly small, inconspicuous, secretive animals, limited in number, both of kinds and of individuals, and restricted in distribution; so it is not surprising that they are in many ways the least familiar of the herptiles. The tropical, burrowing caecilians are completely unknown to most nonbiologists and are little more than names and pictures in a book to many biologists. There are only two modern genera of trachystomes; both are very limited in distribution. The salamanders are somewhat more numerous, and because of their availability in the north temperate zone they have been the subjects of many excellent studies on life history, evolution, and ecology. *Necturus*, of course, is familiar to most comparative anatomy students. Much of our knowledge of how organs are differentiated during development has come from experimental transplanting of tissues on the embryos and early larvae of *Ambystoma* and *Notophthalmus*. Otherwise, the salamanders have been almost ignored by experimental biologists and laymen continually confuse them with lizards.

ORDER GYMNOPHIONA (APODA)

A caecilian is a slim, wormlike creature with no limbs or limb girdles and practically no tail. Most are not more than 240 to 300 mm long. The body is usually divided into a series of segments by folds in the skin that enhance the wormlike appearance. Frequently secondary folds are present between the primary ones. The vent is close to the posterior end of the body on the ventral side; the eyes are minute, without lids, and are buried in the skin; the skull is compact; and the intestine is not differentiated into large and small portions. An unusual sense organ—the tentacle—is present in all caecilians and only in caecilians. It grows forward along the side of the brain and emerges from the skull either at the eye socket or at a point in front of and below the eye. It may not appear externally until after metamorphosis. Its exact function is unknown. Adults lack gills or gill slits. Fertilization is internal and the cloaca of the male is modified to form a protrusible copulatory organ. The more primitive genera have tiny, dermal scales imbedded in the skin; the scales are apparently a heritage from the early, scaled amphibians of the Carboniferous.

Caecilians are primarily forest animals. They have been reported from savanna areas, but only along rivers that are bordered by patches of forests. Members of one family, Typhlonectidae, are river dwellers. All the rest are probably terrestrial and fossorial, living in burrows in damp earth. Snakes, which can hunt caecilians in their burrows, are probably their main predators.

Caecilians are divided into four families.

Family Ichthyophiidae

This family of four genera—*Ichthyophis, Rhinatrema, Caudacaecilia,* and *Epicrionops*—occurs in both Asia and South America. It is characterized by having terrestrial adults that have distinct tails, at least some scales, and from two to four secondary folds on each body segment. *Ichthyophis*, a native of Asia, breeds in the spring. The female prepares a burrow in moist ground close to running water. She coils her body around the 20 or more relatively large-yolked eggs, and guards them zealously from predaceous snakes and lizards. The eggs absorb water and gradually swell until they are about double their original size. The larva at hatching weighs approximately four times more than the newly laid egg. The external gills are lost soon after the larva hatches. The young go through a long aquatic stage before they metamorphose into burrowing, terrestrial adults. Members of the genus *Rhinatrema* of northern South America also lay eggs from which hatch aquatic larvae with external gills.

Family Typhlonectidae

Caecilians included in this family lack tails, scales, and secondary folds. They are the most aquatic of all caecilians. Some species have been taken in the nets of fishermen who were seining streams with gravel or rocky bottoms. The typhlonectids are ovoviviparous and since the fetus has no placental or yolk-sac connection to the uterine wall the source of its energy is an enigma. It has been suggested that it feeds on the soft tissue of the uterine wall. The family is confined to South America.

Family Caecilidae

This is by far the largest of the families of caecilians and is also the most widespread, occurring in Asia, Africa, and South America. The hundred-odd species are divided into twenty-seven genera. The tail is indistinct. The larvae may or may not be aquatic but the adults are always terrestrial. The life histories of most forms are unknown but *Gymnopis* and *Geotrypetes* retain their eggs in the oviducts until they hatch. The wall of the oviduct has compound oil glands and the larvae subsist by eating the wall and its oil droplets. They metamorphose before they are born and thus the newborn young are replicas of the adults.

Family Scolecomorphidae

This small family includes only a single African genus with six species. These caecilians lack secondary folds, scales, and a tail, but have very large tentacles. In the young the eye is covered by a bone of the skull but as the tentacle grows forward it may carry the eye forward so that in some large specimens the eye moves from under the bone. Unlike other caecilians, members of this family lack a stapes. Little is known of the life history, but embryos with curious, large, branching gills have been reported for one species.

ORDER TRACHYSTOMATA

This is another small order that is sharply distinct from other amphibians. Its members are aquatic, permanent larvae that develop few adult characteristics. The maxillaries are tiny and float in connective tissue, and the coracoid ossifies as a separate element. The eyes are tiny. The front legs are minute and the hind legs are lacking entirely. Both jaws lack teeth and are sheathed with horn. The vomerine teeth lack the zone of weakness between crown and pedicel that is found in the teeth of frogs and salamanders. A Jacobson's organ, lungs, and three pairs of external gills are present.

Recent studies of multiple adenosine deaminase enzymes show that these enzymes are very similar in frogs and salamanders but that those of the sirenids differ markedly from the enzymes of the other two orders.

210

FIGURE 13-1
The largest living trachystome, *Siren lacertina*, an abundant
amphibian in the southeastern United States.

The urogenital system of the sirens differs from that of the salamanders.
Glomeruli are well developed in the anterior part of the kidney, the Wolffian duct is relatively straight and inclosed in the kidney, and the ducts from
the posterior part of the kidney show no tendency to fuse into a ureterlike
structure. The male lacks both a Bidder's canal and a vestigial Müllerian
duct. Sirens are unique among vertebrates in the structure of the sperma-
tozoon. It has two axial filaments, each with an undulatory membrane
bordered by a flagellum (see Fig. 6-1).

Courtship and egg deposition have not been observed in any of the three
species but there is evidence that the sirenid reproductive pattern differs
from that of any known salamander. Males lack the cloacal glands that
contribute to the formation of the spermatophore in salamanders with

internal fertilization, and females lack the spermatheca for sperm storage. The absence of these structures has been taken to indicate that sirenids have external fertilization. Among salamanders with external fertilization, the eggs are inclosed together in gelatinous sacs or strings; either the male sheds milt on the egg mass or (in one hynobiid) the eggs are deposited on a sticky spermatophore. The eggs of the sirenid *Pseudobranchus* are deposited singly and attached to the roots of water plants. In Florida, where mats of water hyacinths are the preferred site of deposition, we have never found more than a single egg on any one plant. The eggs may be so widely scattered that fewer than a dozen can be collected in the course of an afternoon. Since a female may have more than 100 ripe eggs in her oviducts, it is obvious that egg deposition must take a considerable amount of time. All salamanders that deposit single eggs that are attached to the substrate have internal fertilization. The pair separates after mating and the female deposits a few eggs each day over a period of a month or so. Either the sirenids practice internal fertilization by some means other than spermatophore deposition or the male accompanies the female during a protracted period and fertilizes each egg singly. There is only one family in this order.

FIGURE 13-2
The egg and newly hatched young of the Narrow-striped Dwarf Siren, *Pseudobranchus striatus axanthus*. The egg is attached to a water hyacinth rootlet. [After Goin.]

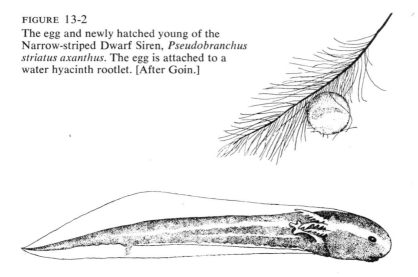

Family Sirenidae

The family Sirenidae includes two recent genera—*Siren* with two species and *Pseudobranchus* with one. It is now restricted to southeastern and central United States and extreme northeastern Mexico. *Pseudobranchus* is a slender little animal that has only three toes on each foot and a single pair of gill slits. Members of the genus *Siren* have four toes on each foot and three gill slits. They are much longer and more heavily built than *Pseudobranchus*. Indeed, *S. lacertina*, which reaches a length of 90 cm, is one of the largest of the living amphibians.

The eggs of *Pseudobranchus* vary from about 7 to 9 mm in diameter. Although the exact developmental time is not known, it must require several weeks since a series of eggs collected in the neural groove stage did not hatch until 17 days later. The newly hatched larvae range from about 14 to 16 mm in total length. There are no balancers, the toes are differentiated, and a well-developed dorsal fin extends from the base of the head to the tip of the tail. There is perhaps less evidence of metamorphosis in the Sirenidae than in any of the perennibranch salamanders, but *Siren* does develop an adult-type skin.

ORDER CAUDATA (URODELA)

As the name of the order indicates, a salamander retains its tail throughout life instead of losing it at metamorphosis as does a frog. The head and trunk regions are distinct. Two pairs of legs and a primitive, poorly developed sternum are present. Fertilization is either external or internal by means of spermatophores. Most salamanders are oviparous. The larvae, which closely resemble the adults, have true teeth in both jaws. Except for the perennibranchs, which are permanent larval types that retain the gills as adults, the lateral line system is lost at metamorphosis.

The salamanders of today do not show the extensive adaptive radiation displayed by many of the other tetrapods. Some are terrestrial, some aquatic, and some may live in either environment. A number are fairly efficient burrowers. Many of the tropical forms have become aboreal, living and reproducing in bromeliads. The majority of the species must stay in or on moist ground, or entirely in water.

The living salamanders are divided into seven families that comprise about 300 species. Many attempts have been made to arrange these families into natural groups (suborders) but there are arguments against every arrangement that has been proposed. Until we know more about the fossil history of the salamanders, it seems better simply to list the families separately.

The first two families are the most primitive of the salamanders and the only ones known to have external fertilization. Other salamanders have a complex of glands in the cloaca that contribute to the formation of the spermatophores. The cryptobranchids and hynobiids possess only one type of cloacal glands and spermatophores are not formed (there is one exception). The eggs are laid in gelatinous sacs.

These two families are also more primitive structurally than the other salamanders. The earliest amphibians resembled the fishes in having many bones in the skull. In the evolution of the group, the tendency has been toward a reduction in the number of bones through loss or fusion. In the cryptobranchids and hynobiids, two of the bones of the lower jaw, the angular and prearticular, are still separate. They are fused in the higher salamanders.

Family Hynobiidae

The Hynobiidae—Asiatic land salamanders—undergo a more complete metamorphosis than the members of the Cryptobranchidae. They develop eyelids, nonlarval teeth, and other metamorphic characters. A stapes develops in the larva and remains separate in the adult, and an operculum is present in adults of only some species. The family is concentrated in eastern Asia and the adjacent islands. There is one widespread, primitive genus, *Hynobius*, whose range extends from Japan to western Asia. (*Hynobius keyserlingi* is sometimes placed in a separate genus, *Salamandrella*.) The other four genera were apparently all independently derived from *Hynobius* and are all included within its geographic range. *Pachypalaminus* and *Batrachuperus* have well-developed, horny epidermal pads on the soles, palms, and digits. *Batrachuperus* differs from both *Hynobius* and *Pachypalaminus* in having the vomerine teeth separated into two small, isolated patches rather than in a V-shaped series. *Ranodon* and *Onychodactylus* are mountain stream forms. Salamanders that live in mountain torrents must keep themselves from being carried downstream by the current. Air-filled lungs make an animal buoyant and increase the chances of its being swept away. Moreover, the water in mountain streams is cool and highly oxygenated. A salamander living in cool water has a low body temperature, and consequently a low metabolic rate. It does not need as much oxygen as a warm water form, and is able to get what it needs from the oxygen-rich water through cutaneous respiration. In such an environment, the disadvantages of lungs outweigh their advantages, and they tend to be reduced or absent. Thus, lungs are small in *Ranodon*, and *Onychodactylus* has become completely lungless, thereby paralleling the plethodontid salamanders of the United States.

Hynobiids have not developed the rather elaborate courtship behavior shown by other salamanders. The male is stimulated to sexual activity by the extruding egg sacs, and his sole response to the female is apparently an attempt to push her away as he fertilizes the eggs. *Batrachuperus karlschmidti* is a common salamander of small mountain streams at elevations of about 1.8 to 4 km in western China. The female attaches the egg case under or to the side of a large stone in flowing water. The end attached to the stone is flat and sticky, and the body of the case is a cylindrical tube that is largest in the middle and smaller toward the transparent free end. The free end is covered with a smooth, rather delicate cuplike cap. This cap is forced off by the movement of the fully developed embryos, which free themselves through the hole thus formed. The individual egg cases contain from 7 to 12 eggs or developing embryos. Since as many as 45 eggs in the same stage of development have been taken from a single specimen, each female presumably deposits 5 or 6 separate egg cases. The larvae are fairly typical salamander stream larvae.

Ranodon sibiricus has been reported to deposit a sticky spermatophore to which the female attaches her eggs.

Family Cryptobranchidae

This family contains the giant salamanders and hellbenders—squat, ungainly, water animals that never completely metamorphose. In contrast to the hynobiids, the adults lack eyelids (as do all larval salamanders) and retain larval teeth. Although they never leave the water, they undergo a partial metamorphosis and adults of both New and Old World forms lose their gills. In the American genus, *Cryptobranchus*, one gill slit remains open, whereas in *Andrias* of eastern Asia all are closed in the adult. The

FIGURE 13-3
The Hellbender of the eastern United States, *Cryptobranchus alleganiensis*.

cryptobranchids also differ from the hynobiids by having flattened skulls from which certain bones (lacrimals and septomaxillaries) have disappeared. *Andrias*, which reaches a length of 160 cm, is the largest of the salamanders, and indeed of the living amphibians. *Cryptobranchus* is about 68 cm long.

Although it was apparently more extensive at one time, this family is now represented by only these two genera. The Hellbender of the eastern United States, *Cryptobranchus alleganiensis*, ranges from New York south to Georgia and west to the Ozarks. *Andrias* comprises two species, *A. japonicus* of Japan and *A. davidianus* of China.

Like the hynobiids, salamanders of the family Cryptobranchidae have external fertilization. *Cryptobranchus* mates in the late summer. The male excavates a nest in the stream bottom beneath a sheltering object, usually a flat rock. He allows females that have not deposited their eggs to enter the nest but drives spent females or other males away. The eggs are laid in long, rosarylike strings, one from each oviduct; these form a tangled mass at the bottom of the nest. As many as 450 eggs may be deposited by a single female and at times several females may lay in a single nest. As the eggs are deposited the male discharges a whitish, cloudy mass that consists of the seminal fluid and the secretions of the cloacal glands. While the eggs develop, the male often lies among them with his head guarding the opening of the nest. The incubation period lasts between 10 and 12 weeks. When the young larvae hatch they are about 30 mm long; they lose their gills when they are about 125 mm long and about 18 months old.

Family Ambystomatidae

These salamanders are usually rather sturdily built, broad-headed animals, small to medium in size. The sides of the body are marked with vertical grooves—the costal grooves—which indicate the position of the ribs. The ambystomatids are more advanced than the hynobiids in having the angular bone fused with the prearticular bone in the lower jaw. The vomers are short and lack posterior processes extending backward below the parasphenoids. Most ambystomatids have a crossbar between the posterior horns of the hyoid apparatus. The condition of the middle ear varies: some have both a separate stapes and separate operculum, in others the operculum is separate but the stapes of the adult fuses to the otic capsule, and in still others the stapes remains separate but the operculum fails to develop. Adults are usually terrestrial. There are three sets of cloacal glands in the male; spermatophores are formed; and fertilization is internal.

The family is found only in North America. The most abundant and widely distributed genus, *Ambystoma*, ranges from eastern to western North America and comprises about 15 species. One of these, *A. tigrinum,* with its various races, is found from coast to coast. Two other genera, *Dicamptodon*

FIGURE 13-4
The Barred Tiger Salamander, *Ambystoma tigrinum mavortium*,
a typical ambystomatid salamander.

and *Rhyacotriton,* each with a single species, are limited in distribution
to western North America. The primitive and partially neotenic *Rhyaco-
siredon*, comprising four species, is found only in the high mountains at
the southern edge of the Mexican plateau. *Dicamptodon* is one of the largest
of the land salamanders, reaching a length of 271 mm. The costal grooves
along the sides, which in many salamanders mark the divisions between
the segments of the trunk musculature (myotomes), are less well defined in
Dicamptodon than in *Ambystoma.* The tail of *Dicamptodon* is more strongly
compressed and thin-edged above. *Rhyacortriton* is a small, mountain
stream genus with vestigial lungs. Many salamanders have an ypsiloid
(Y-shaped) cartilage extending forward from the pelvic girdle on the ventral
side which helps control the shape of the inflated lungs. It is reduced in
Rhyacotriton, as it is in most lungless salamanders.

The breeding habits of *Ambystoma jeffersonianum* are probably typical
of the family. The adults migrate to the breeding ponds in early spring.
Females usually outnumber males and must often bid for attention. There
is a characteristic "Liebespiel" before the spermatophore is deposited. The
female lays the eggs in small, cylindrical masses that contain an average of
16 eggs. These are attached to slender twigs or other objects below the
surface in quiet pools. Since the egg complement of the female may total
over 200, it takes a number of masses to complete egg deposition. The upper
part of the egg is dark brown or black, a pigmentation characteristic of
amphibian eggs that are laid in the open. Under field conditions, the incubation period ranges from about 30 to 45 days. The dark-colored hatchling
is about 12 mm long, and has well-developed balancers; the forelegs are
represented by elongate buds directed backwards, but the hind legs are not
yet developed. The tail fin is continuous with the back fin which extends

almost to the base of the head. Metamorphosis usually takes place two to four months after hatching, and the transforming young may be found from July through September.

Most ambystomatids lay their eggs in water in the early spring, but *Ambystoma opacum* lays its eggs on land in the fall. The female coils around them and protects them. The young, which have all the typical larval characteristics, hatch when winter rains begin and then make their way to the water.

Ambystoma is famous for its tendency toward neoteny. The well-known axolotls of Mexico were long considered a distinct genus of permanent larvae *(Siredon)* until it was found that, by altering their environment, they could be induced to metamorphose into perfectly typical, terrestrial ambystomas.

Family Salamandridae

These are the typical salamanders and newts. They usually metamorphose completely and spend at least part of their lives on land, though an occasional neotenic one is found. A salamandrid has lungs and an ypsiloid cartilage. Costal grooves are not evident. The vomerine teeth are in two long rows, one on either side of the parasphenoid. The stapes fuses to the ear capsule and does not appear as a distinct element in adults.

In ancient times, the European *Salamandra* was credited with being able to live in fire. It is easy to see how such a myth arose. Salamanders sometimes take shelter in crevices in fallen logs. When such a log is thrown on a fire, the animal is roused by the heat and tries to escape, seeming to the unknowing to emerge from the flame. Even today, objects that can withstand a great amount of heat or that are used in producing heat are sometimes called salamanders.

Considerable confusion often arises about the words "salamander," "newt," and "eft." *Salamandra* is a Greek word meaning a lizardlike animal. It is used as the generic name for the common terrestrial tailed amphibians of Europe. As a common noun the word "salamander" is applied generally to all the caudate amphibians, and especially to the more terrestrial salamandroids. "Newt" and "eft" come from the Anglo-Saxon *efete* or *evete*, a word used for both lizards and salamanders. In medieval English this word became "ewt" and finally "a newt" (an ewt). Since the only caudates found in England are moderately aquatic members of the family Salamandridae, newt has come to be used as the common name of the more aquatic salamandrids. Larvae of some of the American newts *(Notophthalmus)* frequently metamorphose into tiny spotted creatures, often bright red or orange in color, that leave the water and live on land for as much as three years. This land creature is called an eft, red eft, or spring lizard. The eft later returns to the water, develops a tail fin and adult coloration, and changes into a sexually mature, aquatic newt. This condition is not fixed, even in a single species, for the larva may metamorphose directly into the aquatic adult. The eft stage is not known in European newts, but the adults frequently spend part of their time on land.

Salamandridae is a widespread family, occupying Europe, eastern Asia, and both sides of North America. It includes about 40 species, most of them Palearctic forms. Best known are members of the genus *Salamandra* and the European newts, *Triturus. Pleurodeles, Triturus,* and *Salamandra* extend into northern Africa and are the only salamanders known from that continent. Only two genera of newts, *Notophthalmus* in the east and *Taricha* in the west, occur in North America.

The salamandrids exhibit a wide variety of rather elaborate courtship patterns. Sight, smell, and touch may all play a part in arousing the female to pick up the spermatophore. During the breeding season, males of some species develop vivid colors and special structures, such as high dorsal crests, which they display before the females. The male may also stimulate the female by the secretions of special hedonic glands, by rubbing, prodding, nipping, or carrying her on a "piggy-back" ride. The male Mountain Newt of Europe *(Euproctus)* actually clasps the female and may place the spermatophore directly in her cloaca. Variation in courtship pattern is probably an important isolating mechanism.

The Red-spotted Newt of eastern North America *(Notophthalmus v. viridescens)* mates in water in the early spring. The male seizes the female and rubs her with his cheeks and chin, thus smearing her with the odorous secretion of his hedonic glands. He then moves a short distance and deposits a spermatophore. She follows, takes it into her cloaca, and some hours later, begins to lay. A female may deposit from 200 to 375 eggs; the period of deposition sometimes lasts for several months. The eggs are laid singly, usually fastened to a leaf or the stem of a small plant, in quiet waters; less often they may be attached to the surface of a stone. The egg is pigmented, with the animal pole varying from light to dark brown. The period of incubation varies from about 20 to 35 days. The larva at hatching is about 7 or 8 mm long and has well-developed balancers, one on each side of the head just below the eye. The front legs are short, blunt buds and the hind legs are undeveloped. As in the larvae of ambystomatids and most other salamanders that breed in quiet waters, a well-developed dorsal keel, continuous with the tail keel, extends nearly to the base of the head. Metamorphosis usually takes place in late summer or early fall, after a larval period of two or three months.

The European *Salamandra salamandra* and *S. atra* retain the eggs in the oviducts for at least a part of the developmental period. The young *S. atra* are born fully metamorphosed. The female *S. salamandra* goes to water to bear the young which are usually born as late-stage larvae. If the embryos of these two species are dissected from the oviduct, they are found to have long, filamentous gills and rudimentary balancers, indicating that these forms evolved from salamanders having pond-type larvae.

Family Amphiumidae

This is the smallest of the families of salamanders. There is only one genus, *Amphiuma*. Three species, all familiarly but inappropriately known as Congo Eels, live today. They are restricted in distribution to the south-eastern United States. Amphiumas are dark-colored, semilarval animals, with long, cylindrical bodies and tiny, useless arms and legs. Large speci-mens of *Amphiuma tridactylum* may be more than 90 cm long. Adults lose the gills though one pair of gill slits remains open. They also show other larval characteristics, such as the lack of eyelids. They have costal grooves and lungs but no ypsiloid cartilage. These salamanders are savage, and large ones can inflict a painful bite.

Courtship takes place in the water. In contrast to most salamanders, in which the female plays a passive role, several females may compete for the attention of a male by rubbing his body with their snouts. The spermato-phore is simple and is apparently transferred directly from the cloaca of the male to the cloaca of the female during a mutual embrace. The eggs are laid in depressions beneath old logs or boards in shallow water, which frequently

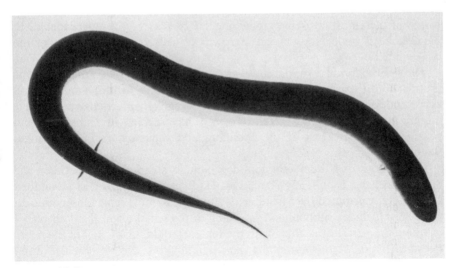

FIGURE 13-5
The Two-toed Amphiuma, *Amphiuma m. means,* of the southeastern United States. Its eellike form has led to the common name Congo Eel.

dries up so that development is finished out of water. The eggs are guarded by the female. The rosarylike strings contain about 150 eggs, each approximately 9 mm in diameter. The newly hatched larvae are from about 60 to 75 mm in total length, of which about 10 mm is tail. They have short, white gills, and are dark brown above and on the sides. Many small, round, light-colored dots are scattered over the surface.

Family Plethodontidae

This is the largest and most successful group of living salamanders. Its members are mostly small to medium in size, and for salamanders, they occupy a wide variety of habitats. Some are aquatic, some terrestrial; some are burrowers, some arboreal. *Typhlomolge*, *Haideotriton*, and *Typhlotriton* are blind, white salamanders that are found in caves and underground waters. The first two are permanent larvae, but *Typhlotriton* metamorphoses and only the adults are blind. All plethodontids are lungless and lack the ypsiloid cartilage. Costal grooves are present. The vomerine teeth lie in patches on the surface of the parasphenoid bone. A small, gland-lined groove —the nasolabial groove—runs from the nostril to the upper lip on each side; it apparently transfers olfactory sense data from the substrate to Jacobson's organ. The middle ear bones of plethodontids differ from those of all other salamanders. The operculum develops first and fills the oval window. The stapes is either absent or present as a slender rod fused to the operculum; it does not enter the oval window. The operculum of other salamanders usually forms at metamorphosis by a pinching off of part of the otic capsule. In plethodontids it develops by an ingrowth of cells from the margin of the oval window.

The Plethodontidae apparently arose in eastern North America. One group (*Bolitoglossa* and related forms) extends into central America and northern South America where it has undergone an extensive adaptive radiation. Another genus, *Hydromantes*, has species in California and southern Europe. Most of the common land salamanders of the United States belong in this family.

Life histories of the Plethodontidae include all gradations between the aquatic and the completely terrestrial. Hedonic glands are well developed and widely distributed over head, body, and tail of most male plethodontids. Courtship includes rubbing and prodding of the female, and the "tail-walk" in which the pair moves along with the female straddling the tail of the male. *Pseudotriton montanus floridanus* is an aquatic type. The eggs are deposited in small groups on tiny rootlets and other submerged objects in cool, muddy springs. In a single clutch, found hanging from rootlets at the edge of an undercut bank, there were 27 eggs in tiny clusters of 2 to 8 eggs. The female stays with the eggs. It is not known how long embryonic development takes,

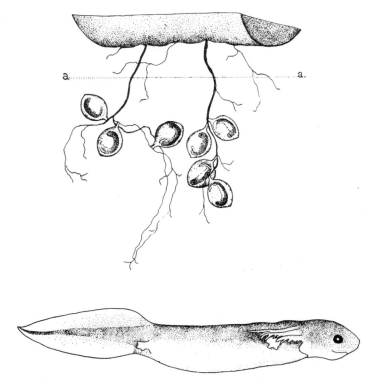

FIGURE 13-6
The eggs and newly hatched larvae of the Rusty Mud Salamander, *Pseudotriton montanus floridanus*; the line *a–a* represents the water line. [After Goin, *Nat. Hist. Misc.*, 1947.]

but the young hatch at lengths of about 12 to 14 mm. Balancers are absent in this species.

The Dusky Salamander, *Desmognathus fuscus*, lays its eggs in small clusters on land in shallow excavations in soft earth, within beds of sphagnum, or beneath stones or logs. These excavations, or nests, are usually located from 30 to 100 cm from the water. Five egg clusters of the southern race contained 9, 11, 14, 15, and 19 eggs. These eggs were approximately 5 by 7 mm in size and each was attached by a short, twisted stalk, about 2 mm long, to a common base along a rootlet. The cluster had the appearance of a bunch of grapes. When the young hatch they are from 16 to 20 mm in total length. They do not go to the water at once, but remain in the nest with the mother and show definite terrestrial adaptations. The hind limbs are longer in proportion to the trunk region than at any time during later development. These young salamanders are thus not merely larvae that have not had a chance to reach the water, but are basically terrestrial salamanders, able to move about in the damp crannies and crevices that lead from the nest to the nearest pool or stream. After one or two weeks, they enter the water to exist as aquatic larvae until they are 7 to 9 months old. They are about 45 mm long when they metamorphose.

The Red-backed Salamander, *Plethodon cinereus,* of the eastern United States, has a typical terrestrial plethodontid life history. The females lay 3 to 12 large, unpigmented eggs in crannies and holes in partly decayed logs. Each egg adheres to the one that preceded it and they form a small mass of eggs that are seemingly contained in a single envelope. This egg cluster is usually attached to the roof of a small cavity in the log. The embryos develop rapidly and soon show large, well-developed, external gills. The gills are lost at hatching and the young have the same form as the adults. They never take up an aquatic larval existence.

The family Plethodontidae is divided into two subfamilies:

Subfamily Desmognathinae.   This subfamily includes three genera, *Desmognathus, Leurognathus,* and *Phaeognathus;* all of them are found in eastern United States. They differ from the other plethodontids in the way they open their mouth. The lower jaw is held in a relatively rigid position by ligaments joining it to the atlas and the mouth is opened by raising the upper jaw and the skull proper. The animals are streamlined with flattened, well-ossified skulls. The hind limbs are larger than the fore limbs. *Leurognathus* is completely aquatic whereas *Phaeognathus* is a large, terrestrial, burrowing form. *Desmognathus* includes the majority of the species; some are semiaquatic, others are terrestrial.

Subfamily Plethodontinae.   This large subfamily contains twenty genera and many species. The mouth is opened by dropping the lower jaw. The group includes aquatic, terrestrial, and arboreal forms and is found from eastern to western North America and from Canada south to Bolivia and Brazil.

FIGURE 13-7
The Slimy Salamander, *Plethodon glutinosus,* a typical plethodontid salamander.

Family Proteidae

This family includes two genera of aquatic salamanders. They are permanent larvae that retain the gills and two pairs of gill openings as adults. They develop lungs and have costal grooves, but lack nasolabial grooves, an ypsiloid cartilage, maxillary bones, and eyelids. The angular and prearticular bones are fused.

Proteus, the blind, white, cave salamander of Europe, has a long, pigmentless body, bright red gills, and skinny appendages that have only three fingers and two toes. Members of the genus *Necturus* have pigmented bodies, stouter limbs, and four toes on each foot; these are the mudpuppies of the eastern United States. There are four species of *Necturus,* and only one of *Proteus.*

Fertilization is internal in the proteids. Little is known of the actual courtship, but in the Mudpuppy, *Necturus maculosus,* mating apparently takes place in the fall and the eggs are laid the following spring. The nests are excavations beneath stones, boards, or other objects, lying in water at depths of 10 to 150 cm. In a lake habitat, nests are found 4 to 8 meters from shore; they have also been found in streams. The eggs are deposited singly and are attached to the undersurface of the object sheltering the nest. Clutch size varies from 18 to 180; clutches that are deposited in streams are usually larger. For example, in five lake nests the average number of eggs per nest was 66, whereas in three stream nests the average number was 107. The eggs, which may be as large as 10 or 11 mm in diameter, hatch after 4 or 5 weeks. The newly hatched larvae are 20 to 25 mm long. There is, of course, no metamorphosis, since these salamanders are perennibranchs.

Proteus usually lays eggs but sometimes retains them in the oviduct where one or two of the young develop. They are born as miniature replicas of the adult. No special modifications, either of the larvae or of the oviducts, are known to accompany this change in life history.

READINGS AND REFERENCES

Bishop, S. C. *Handbook of Salamanders.* Ithaca, N.Y.: Comstock, 1947. (The single, modern, comprehensive account of North American salamanders.)

Boulenger, G. A. *Catalogue of the Batrachia Gradienta s. Caudata and Batrachia Apoda in the Collection of the British Museum.* 2d ed. London: British Museum (Natural History), 1882. (The last comprehensive report of the salamanders of the world, now quite out of date.)

Cope, E. D. "Batrachia of North America." *Bulletin of the United States National Museum,* no. 34, 1889. (The classic in its field. Although the nomenclature is somewhat out of date, this is still a fundamental work on the classification of amphibians.)

Dunn, E. R. *Salamanders of the Family Plethodontidae.* Northampton, Mass.: Smith College, 1926. (Another classic. The preface and introduction should be read by all students of herpetology.)

Francis, E. T. B. *The Anatomy of the Salamander.* Oxford: Oxford University Press, 1934.

Ma, P. F., and J. R. Fisher. "Multiple Adenosine Deaminases in the Amphibia and their Possible Phylogenetic Significance." *Comparative Biochemistry and Physiology,* vol. 27, no. 3, 1968.

Noble, G. K. *Biology of the Amphibia.* New York: McGraw-Hill, 1931.

Salthe, S. N. "Courtship Patterns and the Phylogeny of the Urodeles." *Copeia,* vol. 1967, no. 1, 1967.

Taylor, E. H. *The Caecilians of the World.* Lawrence, Kansas: University of Kansas Press, 1968.

Twitty, V. C. *Of Scientists and Salamanders.* San Francisco: W. H. Freeman and Company, 1966.

Wake, D. B. "Comparative Osteology and Evolution of the Lungless Salamanders, Family Plethodontidae." *Memoirs of the Southern California Academy of Sciences,* vol. 4, 1966.

14

FROGS AND
TOADS

Although frogs and toads are seldom of much interest to amateurs, they are, for a number of very practical reasons, of prime importance to professional zoologists. Their tendency to congregate in large breeding choruses makes it simple to collect them in large numbers. They are easy to maintain alive or to preserve, large enough to be worked on conveniently, small enough not to raise serious problems of storage. Cheap and convenient, they are the laboratory animals *par excellence* of the introductory biology course. They are ideal for studies in experimental embryology, because they lay large numbers of eggs that are easy both to collect and to maintain and hatch in the laboratory, because these eggs are surrounded by a clear jelly rather than an opaque shell, and because most frogs pass through a free-swimming larval stage. They have further potentialities as research animals that are only now beginning to be explored. Ethologists are discovering that territoriality and social hierarchies exist in the frogs. Some of the tropical genera speciate readily and have developed hundreds of distinct forms. These genera may give us clues to mechanisms of speciation that have so far eluded us in studies of more restricted genera.

In the minds of many people, there has been some confusion about the use of the words "frog" and "toad." It should be noted that all members of the Order Anura are collectively and correctly known as frogs. Some frogs, however, may bear the common name "toad," for example, the species of the genus *Bufo* and many others such as the Surinam Toad, *Pipa pipa,* or the Midwife Toad, *Alytes obstetricians.* The term "tree frog" is generally used only for those anurans in which there is a short intercalary cartilage between the ultimate and penultimate phalanges of the digits. These include the members of the Hylidae, Centrolenidae, Rhacophoridae, and Phrynomeridae. The term "tree toad" is best left to the poets.

All Anura are easily recognized, and there is never any doubt whether a particular animal is a member of this order. Every frog lacks a tail in the adult stage; the head is not separated from the trunk by a constricted neck; the legs are well developed, the hind ones being longer than the front ones. Since the great majority of the more primitive frogs are aquatic, it seems logical that the long, muscular hind legs developed first as swimming organs and only later became specialized for jumping. It has also been suggested, though, that the jumping legs evolved to allow animals that lived on the banks of ponds and streams to reach the safety of water in a single bound. Frogs are the most primitive vertebrates to have a middle ear cavity. The tympanic membrane usually lies flush with the surface of the head and is often a very large and obvious structure. Correlated with the development of the ear as a hearing organ is the appearance of a true voice box. Movable eyelids protect the eyes and glands keep the eye moist.

Although the modern frogs are remarkably uniform in gross structure, they have undergone considerable adaptive radiation in their reproductive habits. Fertilization is almost always external. Usually the eggs are laid in water and hatch into aquatic larvae. But all stages are found between this reproductive pattern and true terrestrialism in which the eggs are laid on land and the young hatch as tiny frogs.

Most anurans require moist surroundings and cannot stand prolonged exposure to low humidities, but a few, such as the Giant Toad, *Bufo marinus,* are able to adapt to rather arid conditions. The group as a whole is well adapted to a relatively few niches and in those niches has spread round the world. With the possible exception of the lizards, they are the most widely distributed of the herptiles.

Attempts to divide the Anura into suborders and families have met with many obstacles. As a group, the frogs show a remarkable tendency toward parallelism and convergent evolution. Many characters that were once thought to be indicative of relationship have been shown to have evolved independently in several different lines. No single feature can be relied on to separate the frogs into natural groups. We must consider a complex of characters, reproductive and developmental as well as morphological. Unfortunately, relatively few species have been thoroughly studied anatomically, and the life histories of many are completely unknown. The accumulation of further knowledge will undoubtedly require modifications in the classification given below, and it should in no sense be considered final.

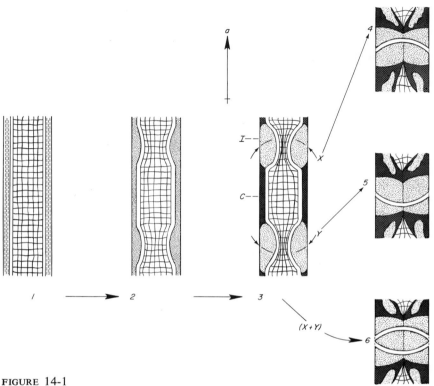

FIGURE 14-1

Vertebral development in anurans: (*1*) early larval stage; (*2*) 22 mm tadpole of *Rana temporaria*; (*3*) *Rana temporaria*, 2 years of age; (*4*) and (*5*) opisthocoelous and procoelous vertebrae, respectively, of *Rana temporaria*, 3–4 years old; (*6*) "free disc" condition of *Megophrys major*, 4 years old. X and Y, invading connective tissue arcs; *(X + Y)*, invasion of a single intervertebral body by two connective tissues arcs; C, ossified cylinder of centrum; I, cartilaginous intervertebral disc. [After Griffiths, 1963.]

One of the most important characters used in the classification of the anurans is the shape of the centrum. In the ontogenetic development of the vertebrae, the centra and intervertebral bodies form as, respectively, thin and thick cylindrical regions alternating along the length of the perichordal tube (see Fig. 14-1). The intervertebral elements chondrify more rapidly than the centra and expand inward to constrict the notochordal tube. Later the centra ossify so that the vertebral column consists of a chain of ossified centra that are joined by undivided intervertebral cartilages. Each intervertebral body is then invaded by an arc of connective tissue. Later a split develops along this arc, dividing the cartilage into two unequal parts, a small cup and a larger ball, which remain attached to their respective centra. Thus the anterior portion forms the posterior face of the centrum in

228

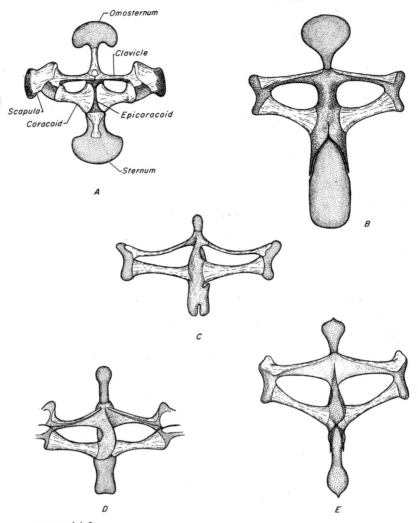

FIGURE 14-2

Pectoral girdle types in frogs: (*A*) the firmisternal girdle in *Rana tigrina*; and variations in the arciferal type, as in (*B*) *Rhinoderma darwini*, (*C*) *Sooglossus seychellensis*, (*D*) *Eleutherodactylus bransfordi*, and (*E*) *Sminthillus limbatus*.

front and the posterior portion forms the anterior face of the centrum behind. The direction of the slope of the arc of invading connective tissue determines which portion forms the cup and which the ball. If the slope runs anteroposteriorly, the ball is formed from the anterior part of the inter-vertebral body, and the cup from the posterior part. This results in a vertebra with a centrum hollowed in front and rounded posteriorly—a procoelous vertebra. If the slope runs the other way, the vertebral centrum is rounded anteriorly and hollowed posteriorly—an opistocoelous vertebra. If the direction of slope alternates in successive intervertebral bodies, a vertebra is formed that has the centrum hollowed at both ends—an amphicoelous vertebra. When the presacral vertebra is amphicoelous and the other vertebrae are procoelous, the vertebral column is said to be diplasiocoelous; that is, it has two types of vertebrae. Sometimes the intervertebral body is invaded by two arcs of connective tissue that slope in opposite directions. This results in the formation of a free intervertebral disc between the centra of successive vertebrae.

Another character that has often been used in the classification of frogs is the construction of the pectoral girdle. If the epicoracoid cartilages on the ventral ends of the two halves of the girdle are separate, with one overlapping the other, the girdle is said to be arciferal (arch-bearing, from the shape of the epicoracoids); if these cartilages are fused, the girdle is said to be firmisternal. However, not all frogs have girdles that can be classified as either arciferal or firmisternal. In some (e.g., the Cuban *Sminthillus*) the epicoracoids may be fused anteriorly, but separate and overlapping posteriorly. Others (e.g., *Sooglossus* of the Seychelles Islands and some species of *Rana*) have the epicoracoids fused anteriorly, then overlapping, then again fused posteriorly.

The structure of the tadpole has also been used in classification. Four basic types of tadpoles have been described. Type I tadpoles have two opercular chambers, each with a spiracle, one on each side, but lack horny mouthparts. Type II tadpoles also lack horny beaks and teeth but have a single opercular chamber with a median, ventral spiracle. Type III tadpoles have an opercular chamber similar to that of Type II but possess horny beaks and teeth. Type IV tadpoles have horny mouthparts and have the spiracular opening of the opercular chamber on the left side. There have been various suggestions as to which of these is primitive and which represent derived states. It seems probable that Type III represents the primitive condition and that the other three types were derived independently from it.

For convenience, we are here dividing the families of frogs into six suborders that are based largely on the structure of the vertebral column.

SUBORDER AMPHICOELA

These are probably the most primitive frogs living today. The name of the suborder is unfortunate, since the vertebrae are not truly amphicoelous. Instead, the intervertebral bodies remain undivided throughout life so that the centra are joined to each other by cartilaginous rings. There are nine presacral vertebrae. Like all frogs, the members of this suborder lack tails, but they show their descent from tailed ancestors by the presence of two tiny "tail-wagging" muscles with formidable names: pyriformis and caudalipuboischiotibialis. Free ribs (not fused to the vertebrae) are present in the adults. The girdle is arciferal in living species. The suborder has only one living family.

Family Ascaphidae

These small grayish frogs are sometimes daintily patterned with pink, brown, and yellow. The family has only two genera: *Leiopelma* in New Zealand and *Ascaphus* in northwestern United States. This widely disjunct distribution is an indication of the antiquity of the group. *Ascaphus* is unique among frogs in that an extension of the cloaca of the male, the so-called "tail," is used as an intromittent organ for the transmission of sperm to the cloaca of the female.

Ascaphus breeds in swift-flowing mountain streams. The male is voiceless. Instead of sitting still and calling for a mate, he swims about on the

FIGURE 14-3
The Tailed Frog, *Ascaphus truei*, of the Pacific coast of North America. The tail is really the everted cloaca, which is used as an intromittent organ.

bottom until he finds one. He then clasps her with a pelvic embrace, humps his body, and maneuvers his extended cloacal appendage into position to thrust into her cloaca. The intromittent transfer prevents the sperm from being swept downstream before they can enter the ova. The eggs are deposited in coils of rosarylike strings which adhere to rocks at the bottom of the stream. In the cold water in which these eggs are deposited, embryonic development is slow and transformation does not occur until the following summer. The tadpoles are Type III.

Leiopelma lays its eggs in damp places on land. The young hatch into terrestrial froglets, though they still have well-developed tails which are resorbed during the first month of growth. The places in which the eggs are deposited are subject to flooding; if the eggs are washed into the water, they are still able to complete normal development. Furthermore, if the embryos are prematurely released from the egg capsules, they can be reared as tadpoles in water.

SUBORDER AGLOSSA

As the name indicates, the members of this suborder lack a definitive tongue. The vertebrae are opisthocoelous. The urostyle is either fused to the sacrum or articulated to it by a single condyle. The shoulder girdle is partly or wholly firmisternal and the ribs, though free in the tadpole, are fused to the vertebrae in the adult. The tadpoles are Type I. The suborder includes only one living family.

Family Pipidae

These are stout-bodied, big-footed, highly aquatic toads. Some forms have the surface of the skin covered with horny tubercles and lateral line sense organs are present in some adults. The Pipidae have the disjunct distribution that is common in ancient and primitive stocks; they are found in Africa south of the Sahara and in South America.

The African pipids, *Xenopus, Hymenochirus,* and *Pseudhymenochirus,* have the tips of the digits undivided and the sternum large. *Xenopus,* the Clawed Frog, is the most primitive and has the widest distribution. All of the South American species are now placed in the single genus, *Pipa.* The tips of the fingers of these frogs are divided to form starlike structures, and the sternum is small.

More striking than the differences in structure that distinguish the species of the two continents are the differences in their life history. The African pipids deposit their eggs in water. The male, at the time of mating, makes a sudden dash at the female and clasps her by the pelvic region. The female expels each egg separately, holding it for a moment by her cloacal lips.

She then grasps a leaf or twig with her hind limbs. The egg is ejected sharply by the cloaca and is propelled along a shallow groove on the venter of the male, past his cloaca and then to the weed, to which it becomes attached.

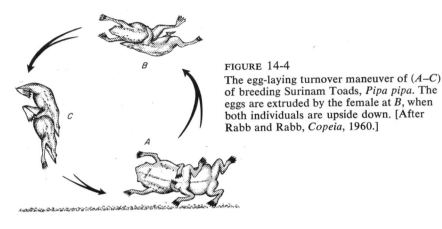

FIGURE 14-4
The egg-laying turnover maneuver of (*A–C*) of breeding Surinam Toads, *Pipa pipa*. The eggs are extruded by the female at *B*, when both individuals are upside down. [After Rabb and Rabb, *Copeia*, 1960.]

The eggs of members of the genus *Pipa* develop in temporary, individual pits that are formed in the soft skin on the back of the female. The method by which the eggs are transferred to the back of the female has been described for a pair of Surinam Toads *(Pipa pipa)* that bred in an aquarium. The male clasped the female inguinally as the pair rested on the bottom of the tank. The dorsal part of the female's vent was pressed against the abdomen of the male. The pair then rose to the top, turning over as they did so, and paused momentarily in an upside down position. At this point three to five eggs were extruded by the female and caught in transverse skin folds on the belly of the male. The two then returned head first to a tilted resting position on the bottom. Fertilization apparently occurred during the descent, and as the frogs righted themselves, the eggs dropped to the back of the female and adhered there. A number of turnovers were required to complete deposition. Fifty-five eggs were thus implanted on the back of the female, and eleven more were lost as they were laid. Eggs of some species of *Pipa* hatch into tadpoles, but others hatch directly into tiny frogs.

SUBORDER OPISTHOCOELA

Members of this suborder have arciferal girdles and opisthocoelous vertebrae. The urostyle is not fused to the sacral vertebra, but articulates with it by one or two condyles. There are eight or nine presacral vertebrae. The suborder includes two recent families.

Family Discoglossidae

Adults of this family of rather primitive frogs have free ribs, a tongue, and movable eyelids. An anterior extension of the procoracoid cartilage splits off to form a separate element, the omosternum. The four genera are all Old World forms. Members of the genus *Bombina,* found in Europe, Asia Minor, and China, are rather flattened, highly aquatic toads. They are brilliantly marked on their underside, black mottled with bright red, orange, yellow, or white. *Discoglossus* is a ranalike frog found in southwestern Europe and northwestern Africa. *Alytes* of western Europe, smaller and more terrestrial, includes the famous Midwife Toad. The completely aquatic *Barbourula* is known only from the Philippines.

FIGURE 14-5
The Korean Fire-bellied Toad, *Bombina orientalis.*

Bombina maxima, the Yellow-bellied Toad, breeds in water in the spring and summer. The male clasps the female with a pelvic embrace. The eggs are laid in small masses which sink to the bottom, or come to rest suspended on vegetation, and hatch in about a week.

The breeding habits of *Bombina* are probably fairly typical for most other members of the family, but the genus *Alytes* shows one of the most remarkable modifications of any frog. The male of the Midwife Toad, *Alytes obstetricans,* calls from a small hole in the ground. Mating takes place on the ground and is apt to last most of the night. The male clasps the female tightly around the head. Just before she extrudes the string of eggs, he moves his hind legs forward so that his heels are together, anterior to and above the cloaca of the female. As the eggs are emitted, he catches the rosarylike strings in his feet and, by stretching his legs backward, delivers from 20 to 60 eggs. He then moves his legs so as to twist the strings around them. He carries the eggs thus for several weeks, until the tadpoles are ready to hatch. Then he makes a brief visit to a pool where no other tadpoles are present. Here the eggs hatch and the Type III tadpoles finish their development.

Family Rhinophrynidae

The Mexican Burrowing Toad *(Rhinophrynus),* the only member of this family, is an egg-shaped, short-legged, toothless creature with a smooth, blotched skin. This aberrant and highly specialized frog is difficult to classify, but is perhaps as closely allied to the discoglossids as to any other group. On the other hand, the tadpoles are Type I, though they differ in details from those of the Pipidae. This may indicate that *Rhinophrynus* should be allied with the pipids. Or the tadpole may have been derived independently from a Type III tadpole. The omosternum is rudimentary and the sternum absent. Free ribs are not present at any stage. The tongue is not attached in front, and the toad is apparently able to protrude it like a mammal rather than to flick it out in normal toad fashion. The foot is specialized for digging —the prehallux is covered with an enormous, cornified "spade" and the single phalanx of the first toe is shovellike. *Rhinophrynus* apparently feeds largely on termites. Amplexus is pelvic, the eggs are deposited in water, and the tadpoles are free-swimming.

SUBORDER ANOMOCOELA

The pelobatid toads and their allies lack ribs at any stage and have true teeth in the upper jaw. The shoulder girdle is arciferal, and an omosternum is present. The vertebrae are either procoelous or amphicoelous with free intervertebral discs. The sacral diapophyses (the vertebral processes to which the pelvic girdle is attached) are expanded. The sacral vertebra is either fused to the urostyle or articulates with it by a single condyle. The tadpoles are Type IV.

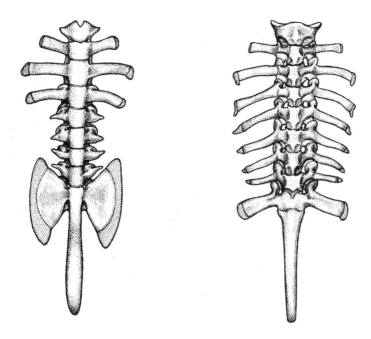

Vertebral columns of (*left*) a frog with expanded sacral diapophyses, *Scaphiopus couchi*; and (*right*) one with rounded sacral diapophyses, *Rana virgatipes*. [After Noble, 1954.]

Family Pelobatidae

Pelobatids usually lay their eggs in open water and pass through a tadpole stage. *Scaphiopus* lays in temporary ponds and eggs and young sometimes show a remarkably rapid rate of development. The eggs may hatch in a day and a half and the tadpoles may transform in fifteen days. *Megophrys longipes* of Malaya is said to lay about a dozen large eggs in humid moss with the young metamorphosing before hatching. There are three subfamilies.

Subfamily Megophryinae. The Megophryinae have free intervertebral discs. The subfamily includes about half a dozen genera and many species, some of them quite large and spectacular. They are found in southeastern Asia, from China to the East Indies. The quaint Nose-horned Frog of the East Indies (*Megophrys nasuta*) has each upper eyelid extended to form a large, thin, pointed "horn," and a flexible projection from the snout, so that the animal looks as though it had three horns. The astragalus and calcaneum, two long bones in the ankle, are separate.

Subfamily Pelobatinae. These are pelobatids with procoelous vertebrae. The Spadefoot Toads of Europe and northern Africa *(Pelobates)* and of North America *(Scaphiopus)* are members of this subfamily. The "spade" is a broad, sharp-edged tubercle on the inner side of the hind foot, with which these burrowing toads can dig a hiding place in short order. Spadefoots resemble the true toads *(Bufo)* in build and are often confused with them; they may, however, be distinguished by their smoother, less warty skins, and vertical, rather than horizontal, pupils. The astragalus and calcaneum are not joined.

Subfamily Pelodytinae. This subfamily includes only one living genus, the slender *Pelodytes* of southwestern Europe and Caucasia. The sacrum is not attached to the urostyle, which has a double condyle. The astragalus and calcaneum are fused into a single bone.

SUBORDER DIPLASIOCOELA

The suborders discussed above are primitive, limited in distribution, and represented today by relatively few species. The suborder Diplasiocoela is an extensive, widespread, successful group, comprising a variety of highly diverse forms. The shoulder girdle is usually firmisternal. Sometimes the epicoracoids are more or less separated, but they are never produced into posteriorly directed epicoracoid horns, as they are in the suborder described next. The vertebral column is either diplasiocoelous or uniformly procoelous. The sacral vertebra articulates with the urostyle by a double condyle (except in the Sooglossinae). Ribs are never present. The suborder includes four families.

Family Ranidae

These are the so-called "true frogs." They are distinguished from the other diplasiocoels by their cylindrical or very slightly dilated sacral diapophyses in conjunction with the absence of an extra element, the intercalary cartilage, between the next to the last and last phalanx of each digit. The tadpole is Type IV. Africa seems to have been the center of differentiation of the ranids; six of the eight subfamilies here recognized are found there, and four of them are found nowhere else. One subfamily, Dendrobatinae, occurs only in tropical America. The genus *Rana,* of the subfamily Raninae, is almost cosmopolitan in distribution.

Very little is known of the reproductive habits of many of the Ranidae, particularly of the African forms. The "typical life history of a frog," as described in most biology books, is almost invariably based upon some species of *Rana.* Most of the species of this genus for which life histories are known conform to the pattern of laying a large number of small-yolked eggs in open water and having a free-swimming tadpole stage. A few African and Oriental species lay their eggs out of water on leaves or stones or in mud burrows, but the egg masses are unmodified and after hatching the tadpoles soon make their way into the water. *Staurois* (subfamily Platymantinae) is a mountain torrent form, with many species in southeastern Asia. The tadpole has a distinctive sucking disc behind the mouth for clinging to rocks in the swift water in which it lives; this abdominal disc is not present in the tadpole of any other frog. Members of the genus *Platymantis* of the East Indies lay large-yolked eggs on land and development is direct. This is also true for the African *Arthroleptis*.

Males of the subfamilies Sooglossinae and Dendrobatinae transport the tadpoles on their backs from the egg-laying site on land to water. The female of *Dendrobates auratus* lays from one to six large-yolked eggs surrounded by an irregular, sticky, gelatinous material with no definite external film. The male either guards or periodically visits the clutch. The eggs hatch in about two weeks and the newly hatched tadpoles wriggle onto the back of the male. He then visits water, the young leave his back, and finish their development as free-swimming tadpoles. They may not metamorphose until six weeks later. Males carrying tadpoles have been found in trees a great distance from any pond or stream, but they may have been about to deposit their offspring in tree holes containing water. Tadpole-carrying males have also been reported in the genera *Phyllobates* and *Colostethus*. A specimen of *Colostethus fuliginosus* has been observed with 25 tadpoles on his back. The larvae of *Sooglossus seychellensis* lack gills and hatch with the hind leg rudiments already developed.

Subfamily Arthroleptinae.   These little African ranids have horizontal pupils and lack vomerine teeth. A few have uniformly procoelous vertebrae. The sternum may be either cartilaginous or bony. (It is usually cartilaginous in the small forms and bony in the larger ones.) Males of a few species of *Arthroleptis,* and of the genera *Schoutedenella* and *Cardioglossa* have the third finger greatly enlarged; it is sometimes as much as two or three times as long as the rest of the hand—a striking secondary sex character unknown among other amphibians. The subfamily includes about six genera and many species; all are little frogs of the forest floor or brushy country.

Subfamily Sooglossinae.   This subfamily includes two genera, *Sooglossus* and *Nesomantis,* both of which are restricted to the Seychelles Islands northeast of Madagascar. The intervertebral body between the sacral vertebra and the urostyle remains undivided. The epicoracoid cartilages are fused anteriorly and posteriorly, and overlap mesially. *Sooglossus* resembles a small arthroleptid externally, whereas *Nesomantis* is more toadlike in build.

Subfamily Dendrobatinae.   This subfamily has three genera: *Dendrobates, Phyllobates,* and *Colostethus.* They are all smooth-skinned little frogs that are found only in tropical America. The vertebral column is uniformly procoelous. (In *Dendrobates* the sacral and presacral vertebrae are fused.) The girdle is firmisternal and the omosternum is frequently bony. A pair of platelike scutes is present on the upper surface of the tip of each digit.

FIGURE 14-7
The Arrow Poison Frog, *Dendrobates auratus,* of Central America.

Subfamily Astylosterninae.   These are West African forest frogs of average size that are characterized by having vertical pupils and a forked, bony omosternum. The subfamily seems to be a natural group of four closely related genera. The terminal phalanges of both fingers and toes of *Nyctibates* are only slightly curved; those of two or more toes of *Scotobleps, Astylosternus,* and *Gampsosteonyx* are bent sharply downward and may pierce the integument to form bony little spines which protrude below the tips of the digits. The functional significance of these modified toes is not known. It may be that they give a sure grip for jumping. The genus *Astylosternus* includes the famous Hairy Frog, *A. robustus,* a species in which the male has a peculiar growth of hairlike processes on the thighs and flanks. These are most fully developed during the breeding season. They are not true hairs, but vascular papillae that presumably aid in respiration.

Subfamily Phrynopsinae.   These small, African frogs, which resemble the members of the genus *Rana,* have horizontal pupils, vomerine teeth, and a cartilaginous unforked omosternum. *Phrynopsis* of Cameroun and Mozambique is big-headed, with elongate, spiky teeth. *Leptodactylodon* of Cameroun is small-headed and has slightly dilated digital discs.

Subfamily Raninae.   These ranids have a bony sternum and pointed, or slightly dilated, digit tips that lack discs on either the upper or lower surfaces. This subfamily includes the genus *Rana* and its close allies. About

FIGURE 14-8
A New Jersey specimen of the Bullfrog of the eastern United States, *Rana catesbeiana.*
This frog has now been introduced into various regions of the world as a potential
source of food.

200 species of *Rana* have been described. Of the major land masses, only Greenland, New Zealand, central and southern Australia, and southern South America lack at least one representative of this vigorous genus. Many are large, and most are colored with soft, muted browns, greens, and yellows. Most of the other genera contain only a few forms and have a limited distribution in Africa or southern Asia.

The largest known frog, *Gigantorana goliath* of Africa, belongs to this subfamily. Some say that this giant of all frogs is really nothing but a large *Rana,* but it differs from that genus in the weak calcification of its epicoracoid cartilages.

Subfamily Petropedetinae.   This small group of African ranids includes two genera: *Petropedetes* and *Arthroleptides.* Both have a pair of glandulo-muscular organs on the upper surface of each digit. In this respect, they are like the South American dendrobatines. The omosternum is bony and either entire or slightly forked posteriorly. The terminal phalanges are T-shaped.

Subfamily Platymantinae.   These are ranids that have the tips of the digits more or less dilated and showing slight indications of the friction pads characteristic of the tree frogs of the following family. The group extends from Africa across the Indo-Australian archipelago to the Solomon and Fiji Islands, and from China to northern Australia. A great many genera and species have been described, but many of them tend to grade into one another and are difficult to define. The subfamily as a whole, though, is a unified group and is apparently the one that gave rise to the family described next.

Family Rhacophoridae

The dozen-odd genera of this family are diplasiocoelous (rarely procoelous) frogs with cylindrical sacral diapophyses. An intercalary cartilage is present between the penultimate and ultimate phalanx of each digit. The extra joint thus provided allows the last phalanx, with its adhesive disc, to be placed flat against the surface regardless of the position of the rest of the foot—an obvious advantage to a climbing form. Most rhacophorids are arboreal, many of them large-disced, brightly colored forms. A few have given up the arboreal habit and returned to the ground, but they retain the adhesive mechanism and intercalary cartilages of their relatives. The tadpoles are Type IV. The family inhabits southern Asia, Japan, the Philippines, the East Indies, Africa, and Madagascar.

Rhacophorids typically lay their eggs in masses of foam on the leaves of plants or other structures above water. The female does most of the work in producing the foam mass. Before the eggs appear, she ejects a small amount of fluid which she beats into a froth by moving her feet mesially and laterally and turning them as she crosses them on the midline. When the foam has been prepared, the eggs and fluid are ejected together. During the egg-laying process, the male is passive, grasping the female under her arm pits and simply holding his body closely applied to her back, his eyes half closed. His pelvic region is bent downward with the cloacal opening near that of the female and the eggs are apparently fertilized as they leave the cloaca of the female. When the egg-laying process is concluded, the female stands on her forelimbs and the male tries to get away from the foam in which the distal ends of his hind legs are buried. The female usually disentangles herself later by moving her legs and body sideways with the help of large, sticky, finger discs.

After the egg-laying, the foam changes color from white to light brown. The eggs are scattered singly or in small groups in the foam mass but are mostly concentrated near the basal part where the foam is attached to the substrate. Just before the eggs hatch, the foam begins to liquefy and the active movement of the fully developed embryos or tadpoles drops them into the water. Sometimes the whole egg-foam mass may be washed into the pool by rain. Tadpoles of a few species sometimes go through their entire development in the nest.

African rhacophorids of the genera *Hyperolius* and *Kassina* lay small, pigmented eggs directly in the water without the foam nest. In contrast, *Rhacophorus microtympanum* of Ceylon lays about 20 large-yolked eggs on land. The female does not produce a foam nest, but remains with the clutch. Development is direct. The embryos bear a striking resemblance to embryos of *Eleutherodactylus,* a member of another suborder described below, which also develop directly on land.

Family Microhylidae

The relationships of this family are obscure. The vertebral column may be either procoelous or diplasiocoelus. The microhylids resemble the ranids in the disposition of their thigh musculature (the semitendinosus remains distinct from the sartorius, its distal tendon passing dorsal to the distal tendon of the gracilis mass). They are firmisternal, but differ from the ranids in having the sacral diapophyses more or less dilated and from the rhacophorids in lacking the intercalary cartilages. Teeth are frequently lacking and the pectoral girdle is reduced. The tadpole is Type II. For the most part, they are dull-colored, inconspicuous little frogs. The family apparently originated in a southern land mass and has spread from there to southeastern Asia, New Guinea, Africa including Madagascar, and to the Americas. It has been divided into seven subfamilies.

Subfamily Dyscophinae. These are microhylids that retain the maxillary and vomerine teeth and have large and undivided vomers on the roof of the mouth. Two genera are found in the Indo-Malayan region and one in Madagascar. The eggs are laid in water and the tadpole is a characteristic microhylid tadpole, with a median spiracle and without horny mandibles and teeth.

Subfamily Cophylinae. The subfamily includes eight procoelous genera, all confined to Madagascar. The vomer, instead of being a single bone, is divided, with the part in front of the choana separated from the remainder. The maxillary and vomerine teeth are reduced. Breeding habits are unknown, but females with large ovarian eggs have been reported in two genera, which indicates that the eggs may be laid on land.

Subfamily Asterophryinae. These are diplasiocoelous, or more rarely procoelous, microhylids that lack teeth and retain the large, primitive, undivided vomer. Four genera containing several dozen species are found in the Papuan region. The eggs are laid on land, development is direct, and the embryo is much like that of *Eleutherodactylus*.

Subfamily Sphenophryninae. These are procoelous species with an undivided, platelike vomer. The teeth are nearly always absent, but some forms have bony, toothlike processes in the vomer. Reproduction is like that of the Asterophryinae. Five genera comprising thirty-six or more species occur in the Papuan region and in Queensland, Borneo, and the Philippines.

Subfamily Microhylinae. In these diplasiocoelous, or more rarely procoelous, species, the vomer is much reduced and is usually confined to the anteromesial border of the choana. Teeth are absent. This is the largest subfamily in the Microhylidae. It ranges from Ceylon and southern and eastern India south to the Celebes, north to Manchuria, and from Argentina to the central United States. Ten of the sixteen genera occur in the Asiatic portion of the range. There seems to be an overlapping of characteristics among the Asiatic and American forms. Reproduction is aquatic.

Subfamily Brevicipitinae. These are diplasiocoelous genera with large, platelike vomers. Teeth are absent. Unlike members of the other subfamilies all the genera retain a complete shoulder girdle. They are confined to southern and eastern Africa and seem to form a closely allied, compact group of four genera.

Breviceps, and probably the other members of the subfamily, lay their eggs in burrows on land. The embryo differs from that of the Asterophryinae in having an operculum and a muscular rather than vascular tail. It hatches at an advanced stage and completes metamorphosis in the nesting site.

FIGURE 14-9
The Mexican Sheep Frog, *Hypopachus cuneus*. The common name refers to the
call of the breeding male. In many of this group the call closely resembles that of a
forlorn lamb.

Subfamily Melanobatrachinae. These procoelous species have small vomers and lack teeth. There are only three genera, confined to the mountainous
regions of southwestern India and Tanganyika in Africa. *Hoplophryne* lays
its eggs in the internodes of bamboos or between the leaves of wild bananas.
The tadpoles are specialized for egg-eating, and superficially resemble those
of the bromeliad-breeding hylas included in the suborder described next.

Family Phrynomeridae

This family of frogs apparently stands in the same relation to the microhylids
as the Rhacophoridae do to the ranids. That is, they seem basically to be
microhylids that have adopted an arboreal existence and have developed
intercalary cartilages. They have diplasiocoelous vertebral columns, dilated
sacral diapophyses, a small vomer restricted to the anteromesial border of
the choana, and a reduced pectoral girdle. The family is entirely African
in distribution and contains only one genus, *Phrynomerus,* with a half-dozen
species widely scattered over Africa south of the Sahara. So far as the life

history is known, there is a free swimming tadpole which resembles that of the typical microhylids.

SUBORDER PROCOELA

A bewildering variety of species, many of them highly specialized, are included in this suborder. The vertebral column is typically procoelous, though occasionally free intervertebral discs are formed. The pectoral girdle is usually arciferal, with posteriorly projecting epicoracoid horns always present. Occasionally the epicoracoids are more or less fused. The urostyle articulates with the sacral vertebra by a double condyle (sometimes urostyle and sacrum are fused). Ribs are never present. The tadpole is Type IV. The suborder is nearly cosmopolitan in distribution and includes six recent families.

Family Bufonidae

These are the true toads, including the familiar, squat-bodied, short-legged bufos of which most of us think when we hear the word "toad." The girdle is arciferal or the epicoracoids are partially fused. The sacral diapophyses are dilated. There is a tendency toward a reduction in the number of vertebrae, apparently resulting from a forward shift of the sacral articulation, with the original sacral vertebra perhaps becoming incorporated into the urostyle. The omosternum is absent (it is present but cartilaginous in *Nectophrynoides*). Maxillary teeth are lacking. Bidder's organ is present in the male.

The genus *Bufo* occurs on all the major land masses of the world except Greenland, Australia, New Guinea, and New Zealand. Two genera, *Dendrophryniscus* and *Oreophrynella,* are found in South America. Twelve other genera of bufonids occur in the tropics and south temperate zone of Africa, southern Asia, and the East Indies.

Bufos typically breed in open water. Amplexus is axillary and the small-yoked eggs are characteristically laid in long strings on the bottom. As with other frogs that lay in open water, the eggs are numerous. The tadpole is short and plump-bodied—the typical polliwog. The tadpole of the Philippine toad, *Ansonia,* which breeds in mountain torrents, is depressed, with a strong, muscular tail, reduced fins, and a suckerlike oral disc. The African *Nectophryne* lays its eggs on land and lacks a free-swimming tadpole stage. Several of the genera (e.g., *Pelophryne* of the Philippines and *Laurentophryne* of Africa) produce relatively few, unpigmented, large-yolked eggs. Little is known of the life histories of these forms; it is possible that they may also lay their eggs away from water. The Brazilian *Dendrophryniscus*

FIGURE 14-10
"Which like the toad, ugly and venomous. . . ."
The American Toad, *Bufo a. americanus.*

brevipollicatus, a rough-skinned little toad of the forest floor, is reported to breed in bromeliads.

The African live-bearing toad, *Nectophrynoides,* diverges most remarkably from the normal anuran reproductive pattern. Breeding takes place on land and fertilization is internal. The male lacks a copulatory organ but the opening of his cloaca is more ventral in position than that of the female. He grasps her in the axillary region and brings his cloacal opening into apposition with hers, thereby transmitting the sperm directly to her cloaca. The young go through their larval development in the oviduct of the mother. The number of eggs is much smaller than in the water-breeding bufos, but even so more than 100 young have been taken from a single female

FIGURE 14-11

Atelopus varius zeteki, the Golden Frog of El Valle de Antón, Panama. El Valle is now a tourist resort and this frog is one of the popular attractions there. Lapel pins modeled after it may be purchased in Panama.

N. vivipara. Although the young are born as fully formed frogs, the embryos still retain some typical tadpole characteristics. However, they lack gills, adhesive organs, labial teeth, and horny beaks. They seem to represent a transitional stage in the evolution toward direct development.

Family Atelopodidae

Members of this family are small to medium-sized frogs, usually brightly and conspicuously marked with red, yellow, and black. The girdle is firmisternal, and the omosternum is lacking. The sacral diapophyses are dilated. Only two genera are included. The variegated toads of the genus *Atelopus* comprise about 25 species, which are widespread in Central and South America. *Brachycephalus* is known only from eastern Brazil. It differs from

FIGURE 14-12
The Mexican Treefrog, *Smilisca baudini*.

Atelopus in the absence of Bidder's organ and in the presence of a broad, dorsal, bony shield, which is confluent with the processes of the second to seventh vertebrae. So far as is known, all atelopodids lay their eggs in water and pass through an aquatic tadpole stage.

Family Hylidae

This vigorous family of tree frogs consists of a group of procoels that have developed intercalary cartilages. They lack Bidder's organ; the sacral diapophyses range from cylindrical to strongly dilated; the pectoral girdle is arciferal; maxillary teeth are present. The family includes 32 genera and several hundred species. With the exception of the genera *Hyla* and *Nyctimystes,* the family is found only in the New World and seems to be centered in the tropics. *Hyla* has spread around the world, but it is missing from most of Africa, from all but the northernmost region of Arabia, from India, and from most of the southern coast of Asia. It is abundant in Australia and New Guinea. *Nyctimystes* occurs only in New Guinea.

Most hylids are arboreal and have enlarged digital discs for climbing, but a few, such as the cricket frogs *(Acris)* of the United States, have become secondarily terrestrial and possess reduced discs. Some hylids have the skin of the head fused to the skull and have developed grotesque, bony casques. Many are spectacularly colored. *Hyla heilprini*, for example, has been described as "pea-green . . . brightly variegated with gold and sky blue."

Various attempts have been made to divide the Hylidae into subfamilies, but none seems satisfactory; we simply do not know enough about the structure of most of the members of this family. Hundreds of species have been described in the genus *Hyla* alone, and the result is an unwieldy and probably unnatural assemblage.

The Hylidae show a wide variety of life history patterns. Most of them, including all those found in the United States, simply lay large numbers of eggs in open water. Some species breed in a still pond or lake, others breed in streams, and the tadpoles are modified accordingly.

Other hylids have more specialized habits. Male Blacksmith Frogs, *H. rosenbergi* and *H. faber,* build basins of mud in or at the edges of pools for the reception of the eggs. The tadpoles have enormous gills with which they cling to the surface film of the small amount of water contained in these mud nests. They metamorphose before leaving the nest.

All of the species of *Hyla* in Jamaica lay their eggs in the small amount of water that is at the base of the leaves of bromeliads. The tadpoles are specialized for feeding on frog eggs or other tadpoles.

Females of the leaf frogs, *Agalychnis, Phyllomedusa,* and *Pachymedusa,* select leaves over water on which to deposit their eggs. The spawning pair moves slowly forward from the tip to the stalk of the leaf, folding it into a nest which is open at both ends. When the tadpoles hatch, they fall through the opening into the water below. They have been placed in a separate subfamily, Phyllomedusinae.

Several South American genera are sometimes placed in a separate subfamily, Hemiphractinae. Structurally, the egg-carrying tree frogs resemble the other hylids, but they differ in that the eggs are carried in a mass on the back of the female. The eggs may be exposed as in *Cryptobatrachus* and *Hemiphractus* or they may be enclosed in a pouch formed by a fold of skin as in *Gastrotheca* and *Amphignathodon.* This pouch is permanent, though reduced during the nonbreeding season, in contrast to the temporary, individual pits in which the eggs of the Surinam Toad *(Pipa)* develop. The young may either leave the pouch as tadpoles and finish their development in water, or metamorphose within the pouch and emerge as young frogs.

Family Leptodactylidae

The Southern Frogs have an omosternum and, usually, a free urostyle. The girdle is typically arciferal, but one subfamily (Rhinodermatinae) has epicoracoids that are partly fused. There is no Bidder's organ; maxillary teeth may or may not be present. The family is widely distributed in the Australian region and in Central and South America, with a few species reaching the southern United States. One genus, *Heleophryne,* occurs in Africa.

Subfamily Leptodactylinae.   Members of this subfamily have an arciferal girdle. The sternum is narrow and may be divided at the posterior end. The sacral diapophyses are usually cylindrical. The group includes many South American frogs; a few species reach the extreme southern United States. It includes the so-called "South American Bullfrog" *(Leptodactylus penta-dactylus)* and, in the West Indies, the Mountain Chicken *(Leptodactylus fallax),* which is large enough to be an important staple in the diet of natives.

The two most abundant genera, *Leptodactylus* and *Eleutherodactylus,* differ in reproductive habits. All species of *Leptodactylus* deposit their eggs in frothy nests that are usually constructed in or near bodies of water. The larvae have very slim bodies and, after hatching, they wriggle through the foam to reach the water. A few species have become more terrestrial. *L. marmoratus* scoops out a small basin in the ground away from water. The eggs and frothy mass are deposited in this basin, which is then covered with mud. The young hatch and pass through the tadpole stage in the nest, escaping after metamorphosis through a tiny hole left in the roof. The eggs of *Eleutherodactylus* are laid on land without protective foam. The developing embryo lacks many typical tadpole characteristics and has a large, flattened, vascular, respiratory tail. The young hatch as fully formed, tiny frogs.

Subfamily Rhinodermatinae.   This is a small group of short-headed little frogs of the neotropics. The epicoracoid cartilages are partly fused and the sternum and omosternum are cartilaginous. The sacral diapophyses are dilated. There is a tendency toward fusion of the vertebrae. The sacral vertebra is particularly apt to fuse to the urostyle. When this happens, it may be hard to determine that the urostyle has the double condyle characteristic of the suborder, since the fused structure of which it is the most conspicuous part terminates in the single condyle of the sacral vertebra. The five genera included in the subfamily have an oddly disjunct distribution: *Sminthillus* is found in Cuba, *Geobatrachus* in Colombia, *Euparkerella* in Peru, *Rhinoderma* in Chile, and *Noblella* in eastern Brazil.

Sminthillus limbatus, which is less than 12 mm long, is perhaps the smallest frog in the world. During the breeding season, the female lays only one large-yolked egg on land. This hatches into a fully formed frog.

The well-known Darwin's Frog, *Rhinoderma,* has one of the most unusual breeding habits of any frog. The female lays 20 to 30 large eggs on land. Several males gather and watch the clutch for 10 to 20 days, until the embryos can be seen moving inside the eggs. Then, over a period of several days, each male picks up several eggs with his tongue and slides them into his vocal pouch. Here the eggs hatch and the young complete the larval stage, emerging only after metamorphosis is complete. Although it lacks a free-living larval period, the developing frog is, for a time, completely tadpole-like, with larval body and tail proportions, a closed operculum, a spiracle, a coiled intestine, and heavy pigmentation. Even the mouth parts are typically larval, although the beaks and labial teeth fail to harden and do not become pigmented.

Subfamily Elosiinae.   This is another small group of leptodactylids, which includes only three genera: *Elosia, Megaelosia,* and *Crossodactylus.* All three are found in eastern Brazil. The terminal phalanges are T-shaped; the omosternum and sternum are cartilaginous; paired, scutelike structures are present on the upper surface of each digit. One genus, *Megaelosia,* has small, bony processes on the mandibular bone of the lower jaw that form pseudo-teeth. They are not true teeth, which are found on the lower jaw of frogs only in *Amphignathodon,* a member of the family Hylidae.

Little is known of the breeding habits of these frogs. They live in the vicinity of mountain streams, and the tadpoles may be found hiding on the bottom, under overhanging margins of pools, or behind large fragments of rock.

Subfamily Heleophryninae.   The few species of leptodactylids known from Africa belong to a single genus, *Heleophryne.* The terminal phalanges are T-shaped, maxillary teeth are present, the omosternum is lacking, and the sacral diapophyses are moderately dilated. The tadpoles are found in mountain torrents and are modified for that type of habitat, as are those of the preceding subfamily.

Subfamilies Cycloraninae and Myobatrachinae.   The many leptodactylids of the Australian region are placed in two separate, though closely related, subfamilies. They have dilated sacral diapophyses and a broad, usually cartilaginous, sternum. They have not been clearly differentiated from the South American forms, and there is need for a thoroughly anatomical study comparing them to the other leptodactylids. They have been given sub-familial rank, apparently because they are so widely separated geographi-

FIGURE 14-13
The Brazilian Horned Toad, *Ceratophrys cornuta*. These frogs use their
huge mouths for catching other frogs, on which they feed.

cally from other members of the family, but this is not a valid criterion on which to base classification.

The two subfamilies differ mainly in the structure of the mouth parts. The Cycloraninae have large tongues, and the covers and vomerine teeth are well developed; the Myobatrachinae have small tongues, and the vomers and vomerine teeth are reduced or absent.

Life history data are lacking for many of the species. Some of the Cycloraninae produce foam nests similar to those of *Leptodactylus*. The Marsh Frog, *Heleioporus eyrei,* lays its eggs in a frothy mass underground. Development proceeds within the egg until the external gills have been lost and the operculum has developed. Hatching takes place when the nest is flooded. The Marbled Frog, *Limnodynastes tasmaniensis,* lays small eggs in a foam nest floating on any available water supply. These eggs hatch in about 48 hours. *Kyarranus* lays its eggs in sphagnum or moist earth and apparently lacks a free-swimming tadpole.

Family Ceratophryidae

At least seven genera of toadlike, wide-mouthed, rather bizarre forms are included in this small family. They are restricted to South America. Their toadlike mien, enormous mouths, and the absence of a bony style on the sternum distinguish them from their close relatives, the leptodactylids.

There is not much known of the life histories of most of the members of this family. Frogs of the genus *Ceratophrys* are extremely belligerent animals that snap and bite viciously at any thing that approaches. *Ceratophrys* and *Lepidobatrachus* are known to feed on frogs and it is not unlikely that members of the other genera do also.

Family Pseudidae

This family includes two small genera of highly aquatic South American frogs. An extra phalanx is present in each of the digits. This increase in phalangeal formula is apparently an adaptation for swimming, similar to that of aquatic mammals. The extra element is a different structure, and serves a different function from the one found in the foot of the various groups of tree frogs. The sacral diapophyses are cylindrical, maxillary teeth are present, and the thumbs are opposable. Reproduction is aquatic; the eggs are laid in a frothy mass. *Pseudis paradoxus* is remarkable for the large size of the tadpoles, which may be more than 250 mm long (the adults are only about 65 mm long).

Family Centrolenidae

This small family of arboreal frogs is apparently closely related to the Leptodactylidae. As in other families of tree frogs, intercalary cartilages are present. The terminal phalanges are T-shaped, the sacral diapophyses are dilated, and the astragalus and calcaneum are fused into a single, slender bone. The pectoral girdle is arciferal, and there is no omosternum. Most centrolenids are small, bright green frogs. The males of *Centrolene* and of some species of *Centrolenella* have a long, sharp, bony spine growing from the proximal end of the humerus. The spine pierces the skin of the upper arm and fits into a pocket in the forearm. *Teratohyla* has a sharp spine projecting from the pollex rudiment. The function of these spines is completely unknown. Breeding habits are not well known. Eggs are deposited in disc-like masses on the undersides of green leaves above running water, and are apparently guarded by the male. The tadpoles fall into the water after hatching. This family, which includes three or four genera, is found from Mexico to Paraguay.

READINGS AND REFERENCES

Ahl, E. *Amphibia: Anura III, Polypedatidae.* Das Tierreich, pt. 55. Berlin and Leipzig: Walter de Gruyter, 1931.

Boulenger, G. A. *Catalogue of the Batrachia Salientia s. Ecaudata in the Collections of the British Museum.* 2d ed. London: British Museum (Natural History), 1882. (Modern herpetology really begins with the British Museum Catalogues.)

Cochran, D. M. *Living Amphibians of the World.* New York: Hanover, 1961.

Cope, E. D. "Batrachia of North America." *Bulletin of the United States National Museum,* no. 34, 1889. (The cornerstone of modern amphibian classification.)

Griffiths, I. "The phylogenetic status of the Sooglossinae." *Annals and Magazine of Natural History,* ser. 13, vol. 2, no. 22, 1959. (In this and other papers, Griffiths applies modern techniques of comparative anatomy to some basic problems of the classification of frogs.)

———. "The Phylogeny of the Salientia." *Biological Reviews of the Cambridge Philosophical Society,* vol. 38, 1963.

Kluge, A. G., and J. S. Farris. "Quantitative Phyletics and the Evolution of Anurans." *Systematic Zoology,* vol. 18, no. 1, 1969.

Mertens, R. *The World of Amphibians and Reptiles.* London: George G. Harrap, 1960.

Neiden, F. *Amphibia: Anura I, Subordo Aglossa und Phaneroglossa.* Das Tierreich, pt. 46. Berlin and Leipzig: Walter de Gruyter, 1923.

———. *Amphibia: Anura II, Engystomatidae.* Das Tierreich, pt. 49. Berlin and Leipzig: Walter de Gruyter, 1926.

Noble, G. K. *Biology of the Amphibia.* New York: McGraw-Hill, 1931.

Reig, O. A. "Propasiciones para una nueva macrosystemática de los anuros." *Physis,* vol. 21, no. 60, 1958. (The classification used in this chapter is based in part on that proposed by Reig.)

Tihen, J. "Evolutionary Trends in Frogs." *American Zoologist,* vol. 5, 1965.

15

TURTLES

Salamanders are sometimes mistaken for lizards, and lizards for snakes, but a turtle can never be mistaken for anything else. The short, wide body is encased in a protective armor, the shell, which is composed of a dorsal carapace and a ventral plastron. The carapace is formed of dermal bones that are usually fused to each other and to the underlying vertebrae and ribs, and that are typically covered with large, epidermal scales, the laminae (sing. lamina). The bones of the plastron apparently evolved from parts of the shoulder girdle and from gastralia, and are also covered with laminae. The laminae do not correspond in shape or size to the underlying dermal bones. Usually a bony bridge on either side, formed by an extension of the plastron, connects upper and lower shell (see Fig. 15–1).

All turtles lack teeth. Instead, each jaw is usually covered with a horny sheath, the beak, which has a sharp cutting edge, the tomium. The skull is anapsid. The shoulder girdle is unique in that it lies beneath the ribs. The vent is a longitudinal slit and the glans penis is single except in the families Trionychidae and Carettochelyidae, in which it is multilobed. The tail is short.

The confusion arising from the varied usages of the words "turtle," "tortoise," and "terrapin," is even worse than that between "frog" and "toad." Thus, although the terrestrial forms are usually called tortoises, the members of the terrestrial genus *Terrapene* are called neither terrapins nor

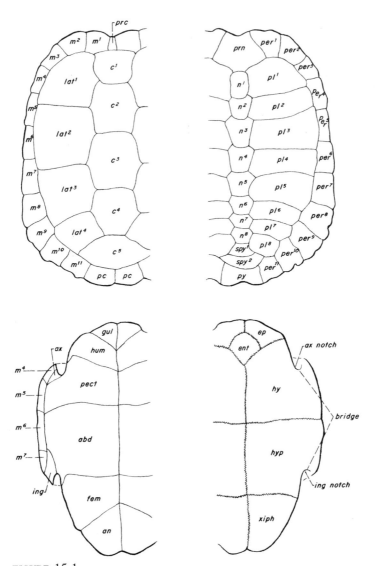

FIGURE 15-1

Epidermal shell (*left*) and bony shell (*right*) of an emydid turtle
(*Chrysemys s. scripta*). Epidermal laminae: *prc*, precentral; *m*,
marginals; *lat*, laterals; *c*, centrals; *pc*, postcentrals; *gul*, gular; *hum*,
humeral; *pect*, pectoral; *abd*, abdominal; *fem*, femoral; *an*, anal;
ax, axillary; *ing*, inguinal. Shell bones: *prn*, proneural; *per*, peripherals;
pl, pleurals; *n*, neurals; *spy*, suprapygals; *py*, pygal; *ep*, epiplastron;
ent, entoplastron; *hy*, hyoplastron; *hyp*, hypoplastron; *xiph*,
xiphyplastron; *ax notch*, axillary notch; *ing notch*, inguinal notch.
Some turtles have a series of laminae known as inframarginals that
separate the axillary and inguinal laminae. [From Carr, *Handbook
of Turtles*, Comstock, 1952.]

tortoises, but Box Turtles. On the other hand, tortoiseshell is not obtained from the terrestrial tortoises but from the marine Hawksbill Turtle. All members of the order may properly be called "turtles." We shall restrict the name "tortoise" to the members of the subfamily Testudininae—the Giant Tortoises and their allies—and use "terrapin," not for any systematic group, but for such small to medium-sized, more or less aquatic, hard-shelled turtles as are commonly used for food.

Many modern turtles are semiaquatic marsh dwellers and it seems probable that this has been their characteristic mode of existence throughout the course of their evolutionary history. There have been three major adaptive trends away from this mode: the tortoises have become terrestrial, other turtles are truly aquatic or even marine, and still others have adopted a bottom-dwelling existence. It is a curious anomaly that, although the primary turtle adaptation was the development of a heavy, bony box, these three adaptive trends have all involved a reduction of the shell.

Many fossil turtles are noteworthy for the thickness of the carapace bones, but such a heavy shell would be too much of a burden for a creature moving overland. Modern turtles, and particularly the larger terrestrial tortoises, have met the problem of the weight-to-volume ratio, not through a reduction of the number of bones of the shell, but through a reduction in the thickness of the individual bones. Many land turtles also developed a high, domed shell, perhaps in part as a protection against gnawing predators that might more easily crack a flat one. The African Pancake Tortoise, *Malacochersus tornieri,* which lives in rocky terrain, has such thin bones in the flattened carapace that the shell is flexible. The animal can squeeze into narrow crevices or between boulders and when it is in place it inflates its body with air, making it almost impossible to remove.

FIGURE 15-2
The African Pancake Tortoise, *Malacochersus tornieri*, which is adapted for living in rocky crevices.

Except for perhaps the sea snakes, the most aquatic of all living reptiles are the sea turtles, which are adapted for swift movement through the water. The Leatherback, *Dermochelys,* has undergone an extensive loss of shell bones; its carapace is made of a mosaic of small dermal elements. Other marine turtles have a reduced plastron and the embryonic gaps between the lower ends of the ribs sometimes persist even to maturity. The sea turtles have also developed an efficient swimming stroke. The forefeet have become flippers and are moved with an up-and-down beat similar to the motion of a bird's wing in flight. Tortoises are traditionally "Slow-and-Solid" but marine turtles are among the swifter of the tetrapods with a probable top swimming speed of 32 km per hour.

Other turtles, such as the snappers, softshells, and the bizarre South American side-necked turtle, *Chelys fimbriata,* are adapted for bottom dwelling. The plastron of the snappers is greatly reduced. The softshells lack the laminae entirely. The shell is covered instead with an undivided, leathery skin whose edges conform to the bottom contour as the animal lies hidden under silt or sand.

Of all the reptiles, turtles are of the greatest direct economic benefit to man. They are probably eaten everywhere they occur. Not only do they provide the gourmet with green turtle soup and terrapin stew, but they and their eggs are a major source of protein in many parts of the world where meat is hard to get. The dorsal laminae of the marine Hawksbill Turtle are the source of the beautifully marked tortoiseshell, used for many centuries to make jewelry and other *objets d'art.*

Turtles are conservative in their breeding habits. All species are oviparous and even the marine forms must come to shore to bury their eggs above the high tide mark.

The order Testudinata, a small group today, comprising about 335 species, is divided into two suborders—Cryptodira and Pleurodira. Cryptodira has five superfamilies. That such a relatively small number of genera (about 65) can be separated into so many sharply distinct groups is indicative of the extreme antiquity of the order.

SUBORDER PLEURODIRA

The turtles of this suborder are known as side-necked turtles because they withdraw their heads by bending their necks laterally instead of vertically as do the cryptodires. The cervical vertebrae have rather high spines posteriorly and well-developed transverse processes for the insertion of the muscles that bend the neck. Their central articulations are well developed, but are always single. The pelvis is fused to the plastron and sutured to the carapace. The temporal roof may be emarginate (notched) from behind and is usually more or less emarginate from below. A pair of mesoplastra (small bones lying between the anterior hyoplastra and posterior hypoplastra) are sometimes present.

These aquatic turtles are found today only in the southern continents— South America, Africa including Madagascar, and Australia. The suborder is divided into two families.

Family Pelomedusidae

These turtles are able to tuck the head and neck into the shell so that the neck is concealed; hence they are sometimes called hidden-necked turtles. The skull is emarginate behind, nasal bones are absent, and the second cervical is biconvex. Mesoplastra are present. The family includes three Recent genera: the African *Pelomedusa* and *Pelusios* are moderate-sized forms, with maximum shell lengths of around 30 cm; *Podocnemis* of South America and Madagascar is larger, attaining a shell length of 80 cm.

FIGURE 15-3
The Marsh Side-necked Turtle of Africa, *Pelomedusa subrufa*.

All of these turtles are aquatic, although *Pelomedusa* may wander freely overland. It is reported to aestivate during the African dry season. *Pelomedusa* is said to be carnivorous; *Podocnemis* is largely herbivorous.

Podocnemis expansa breeds on sandy islands midstream in the Orinoco and Amazon rivers. The females crowd together on the tiny islands, each digging a hole 50 to 60 cm deep and depositing 80 to 200 eggs. For years, natives of these regions have conducted highly organized egg hunts, gathering them by the millions and mashing them to extract the oil. The natives also prey on the hatchlings as they emerge and hunt for the adults in the rivers. Only a species with a very high biotic potential could long survive such heavy predation, and the turtles are no longer as numerous as they once were. The population will probably continue to decline until it is no longer profitable to hunt for them.

Family Chelidae

The head of any of the turtles in this family can be more or less withdrawn under the margin of the carapace but the neck remains exposed. The skull is little emarginate from behind, nasal bones are usually present, the fifth and eighth cervical vertebrae are biconvex, and mesoplastra are lacking. The jaw is usually long, slender, and weak. Because the neck is often longer than the carapace, these turtles are sometimes known as snake-necked turtles. The carapace is from about 15 to 40 cm long.

FIGURE 15-4
A South American snake-necked turtle of the
family Chelidae, *Platemys platycephala.*

There are ten genera of Chelidae: four are found in Australia and New Guinea, six in South America.

One of the most bizarre creatures that ever lived is the South American Matamata *(Chelys fimbriata)*. Its snout is drawn out into a snorkel-like proboscis with the nostrils at the tip, and the tiny eyes are placed far forward in the head. The mouth is very large, reaching back to the region of the ears, but the jaws are weak. The skin of the sides, the lower part of the head, and the long, thick neck is fringed and frayed. The carapace is rough and ridged and has three broad keels.

The female of the Long-necked Turtle of Australia *(Chelodina longicollis)* scoops a circular hole in the ground and deposits as many as 20 elongate, white eggs. They are laid in November or December and the young hatch in February or March.

SUBORDER CRYPTODIRA

This suborder includes almost all the modern turtles. It is distributed on all continents though only marine forms reach the shores of Australia. The skull roof is usually very emarginate from behind. There are never any mesoplastra, and the pelvis is never fused to the plastron. The cervical vertebrae are distinctive. The postzygapophyses are set wide apart. The central articulations are well developed, broad, and typically double on the posterior cervicals. The posterior cervical spines are low and the transverse processes are almost absent. The neck is more or less retractable in a vertical position, bending in a sigmoid curve as the head is drawn into the shell. There are five natural groups within the suborder Cryptodira.

SUPERFAMILY TESTUDINOIDEA

The four families included in this superfamily are the modern amphibious and terrestrial turtles. The shell is usually complete and epidermal laminae are always present. The limbs are not modified into paddles as they are in the sea turtles. There are one or two biconvex centra in the neck, and two or three of the neck joints are usually double.

Family Dermatemydidae

Of this rather primitive family, only one genus, with a single species, is living today. This is *Dermatemys mawi,* found from Vera Cruz, Mexico, to Guatemala and Honduras. It is a rather flat, aquatic turtle with a very short tail; it is about 45 cm long. The plastron is well developed, not cruciform in shape, and is joined by sutures to the carapace. A row of inframarginal laminae on the bridge seems indicative of its primitive position. There is only one biconvex centrum in the neck and the eighth centrum is doubly concave in front. The tenth dorsal rib is not fused to the pleural. The alveolar surface of the maxillary bone (the crushing surface of the upper jaw) is broad and ridged.

In countries where these turtles live, natives fish for them in muddy backwaters and oxbow lakes, and sometimes sell them in the markets. The species

FIGURE 15-5
The "Jicotea" of Central America, *Dermatemys mawi.*

is extremely aquatic and has difficulty moving on land. In fact on land it seems unable or unwilling to even hold its head off the ground, a phenomenon that can be noted in most photographs of these turtles. Little is known of its breeding habits other than that it lays about 20 eggs in the fall when the rivers are flooded; the eggs are then covered with decaying vegetation.

Family Chelydridae

This small, New World family includes one of the largest freshwater turtles, the massive Alligator Snapping Turtle *(Macrochelys temmincki).* The proneural bone has costiform processes which extend beneath the marginals. The tenth dorsal vertebra is usually without ribs. The alveolar surface of the maxilla is typically broad and without ridges. There is one biconvex centrum in the neck and the eighth centrum is usually doubly concave in front.

The snapping turtles are large, with big heads, powerful jaws, long tails, small plastrons, and ugly tempers. The carapace is rough and ridged; the small plastron is cruciform and loosely joined to the carapace by a narrow ridge. An entoplastron is present; there are 11 marginal laminae on each side. The feet are webbed, and the toes are long. The Alligator Snapping Turtle of the southeastern United States may reach a weight of 100 kg. The smaller Common Snapping Turtle *(Chelydra),* which is found from southern Canada to northern South America, seldom weighs more than 25 kg. Both forms are aquatic, though Common Snappers sometimes wander overland, and they eat both plants and animals.

FIGURE 15-6

The Alligator Snapping Turtle, *Macrochelys temmincki*, of the southeastern United States. Note the bait on the floor of the open mouth.

The breeding habits of the Common Snapper *(Chelydra serpentina)* are highly variable. Mating occurs between late April and November. During copulation, the male holds his position atop the female by clinging to the under edge of her shell with the claws of all four feet. He then twists his tail upward and manipulates it until it establishes contact between the vents. The eggs are most commonly laid in June, but nesting has been reported from May to October. The nest is dug at various distances from the water, its site is apparently determined not only by the nature of the soil but also by the whim of the female. Nests have been found at distances of 1 to 25 meters from water; they are about 10 to 18 cm deep, and vary widely in form and in manner of excavation. Sometimes, at least, the cavity is definitely flask-shaped, and the excavation slants downward. The female digs by alternately working her hind feet. She guides the eggs to the bottom with one foot and covers the nest before leaving. The number of eggs deposited varies from 8 to 77 (the latter in a single nest in Manitoba). The eggs may vary from about 25 to 33 mm in diameter, and although often spherical, they are sometimes slightly elongated. They hatch in about three months.

Family Kinosternidae

The four genera of turtles in this family differ from those in preceding ones by consistently having 10 rather than 11 marginals on each side. In the two Central American genera, *Staurotypus* and *Claudius,* an entoplastron is present while it is absent in the North American genera *Kinosternon* and *Sternothaerus,* a character almost unique among turtles.

Claudius, Kinosternon, and *Sternothaerus* are rather small turtles but *Staurotypus* may attain a length of 375 mm.

The anterior and posterior parts of the plastron of *Kinosternon* are nearly equal and are hinged and movable on the central part. The plastron of *Sternothaerus* is smaller, and the anterior lobe is shorter than the posterior and is scarcely movable. *Sternothaerus* is restricted to eastern North America north of Mexico; *Kinosternon* is found from the United States to northern South America.

These turtles are largely carnivorous or carrion feeders, but they also take some plant food. Although highly aquatic, they sometimes travel overland and most species are fond of basking in the sun.

The Eastern Mud Turtle *(Kinosternon s. subrubrum)* usually nests in the early summer, but the nesting sometimes lasts until September. The female searches until she finds a suitable site. She digs with her forefeet, thrusting out the dirt laterally until she is almost concealed, then she turns around and completes the nest with her hind feet. While she is digging with the hind feet, and while laying, only her head is visible above the ground. After two to five eggs have been deposited, the turtle crawls out and may return directly to the water, or may make a slight effort to conceal the nest cavity by levelling and scratching the earth around the site. The completed nest is a semicircular cavity 75 to 130 mm deep and inclined at an angle of about 30 degrees. The cavity extends slightly beyond the eggs; the soil immediately around them is firmly packed, indicating that the turtle covers them carefully even though she makes no effort to conceal the entrance to the nest. Rains, pounding on the loose, sandy soil, soon obliterate all sign of it. Apparently, the eggs sometimes over-winter and hatch the next spring, for hatching eggs have been plowed up in April. In the laboratory, hatchlings have emerged in late September.

Little is known of the life history of the Central American genera, *Claudius* and *Staurotypus*.

Family Testudinidae

This is the one flourishing group of modern chelonians. Most of the familiar pond and swamp turtles, as well as the terrestrial tortoises, are members of this family. The carapace and plastron are unreduced and the bridge between them is well developed. The proneural is without lateral costiform processes. There are usually two biconvex centra in the neck and the eighth centrum is typically doubly convex in front. The alveolar surface of the upper jaw may be broad or narrow, and it frequently bears one or more ridges. The family is divided into three subfamilies.

Subfamily Platysterninae. *Platysternon megacephalum,* the only member of this subfamily, is a rather small (carapace length about 150 mm) turtle with a big head, long tail, strongly hooked jaws, and depressed shell. The plastron is connected to the carapace by ligaments, and the inframarginal row is complete. There are 11 marginals on each side. The temporal region of the skull is almost completely roofed. The digits have three phalanges, which are not quite fully webbed.

These turtles inhabit mountain streams in southeastern Asia. Rather surprisingly, they are agile climbers and may be found in trees or on rocks, hunting for food or sunning themselves. A captive specimen ate snails and worms. The female lays only two eggs at a time.

Subfamily Emydinae.   This most extensive group of living turtles includes both aquatic and terrestrial forms. Most are medium sized, but some are among the largest of the nonmarine turtles. The temporal region of the skull is not roofed. The middle digit usually has three phalanges and the toes are usually more or less webbed. The inframarginal series of laminae is never complete; there are 11 marginals on each side.

The Emydinae includes about 25 genera; representatives are found on all continents except Australia and Africa south of the Sahara. The richest and most varied fauna is that of eastern and southeastern Asia, where there are 17 or more genera. Europe has only two genera, *Emys* and *Clemmys.* The subfamily is well represented in eastern North America, where it includes such well-known forms as the terrestrial box turtles *(Terrapene)* whose plastron is hinged so that the shell can be closed completely, and the succulent Diamondback Terrapin *(Malaclemys).*

Food habits within the group are varied. Some species are strictly herbivorous, others are carnivorous, and still others seem to prefer a mixed diet.

Some of the emydines have developed elaborate courtship patterns (see p. 189). Most species lay relatively few (less than 12) rather large eggs, but some of the larger forms may deposit 30 or more at a time. *Batagur baska* is a large (shell length of nearly 60 cm) aquatic species found in estuaries, slow-flowing rivers, and canals of southeastern Asia. Around the mouth of the Irrawaddy River, egg laying takes place during January and February. Every afternoon the turtles come ashore, gather in large herds, and sun themselves on the sand. At night the females dig holes 45 to 60 cm deep in the sand above the high tide mark and deposit 10 to 30 eggs each about 75 mm long. Each female lays between 50 and 60 eggs, in three batches, over a period of about six weeks. The incubation period is about 70 days.

Subfamily Testudininae.   These are the true land tortoises, including the lumbering giants of the Galápagos and Seychelles Islands. The feet are club-shaped, short and broad, and do not have more than two phalanges in any digit. The toes are completely unwebbed. The cylindrical and columnar hind legs are like those of an elephant. The skull roof is incomplete posteriorly; the inframarginal series is incomplete; there are 10 or 11 marginals on each side. Usually the shell is high and dome-shaped. Not all tortoises are large—the little *Testudo leithi* reaches a shell length of only 120 mm. The largest testudine recorded, a specimen from Aldabra Island, measured 1400 mm in straightline shell length and weighed 254 kg.

FIGURE 15-7
Hermann's Tortoise, *Testudo hermanni*, of southern Europe. Breeding males of this and other species of *Testudo* exercise their voices in the spring.

Nine recent genera are recognized in this subfamily. By far the largest and most widely spread is the genus *Testudo,* found in southern Europe, Asia, and Africa, as well as on islands of the Indian Ocean. North America has only one genus, the Gopher Tortoise, *Gopherus,* of southern United States and Mexico. *Geochelone* is found in both the Old World and the New World. The remaining genera are all confined to Africa and Madagascar.

In general, each tortoise has an individual resting place—a burrow it has dug, a cranny in a rock, or a sheltered nook under a plant—from which it emerges to graze during the day or, in hot weather, in the cool of the evening. Tortoises are largely herbivorous, but *Testudo graeca* has been reported to take insects, molluscs, and worms. One captive Galápagos tortoise was seen to catch two rats and a pigeon, and a number of zoo specimens from the same area have eaten raw meat greedily. Tortoises that live in desert areas, or on rocky islets where no fresh water is available, are able to get along without drinking; other species drink copiously and enter water freely.

Courtship of tortoises seems to consist mainly of the male pursuing the female and butting her with his shell. Copulating males are surprisingly vocal —the "voice" has been described for various species as "a muffled, whistling cry," "a peculiar grunting noise," and "the intermittent winding up of a metal spring." In spite of their relatively large size, most female tortoises lay few eggs in a single clutch, usually less than seven. Some species may lay only a single egg per clutch. As with most other turtles, the eggs are buried in a hole which the female digs with her hind legs.

SUPERFAMILY CHELONIOIDEA

These are large, sometimes enormous, marine turtles which have epidermal laminae, a heart-shaped carapace, limbs modified as paddles, and a skull roof more or less complete posteriorly. There is only one biconvex vertebra in the neck, and the head cannot be withdrawn completely into the shell. The superfamily contains only one living family.

Family Cheloniidae

The family includes four living genera: *Caretta,* the massive Loggerheads which may weigh as much as 400 kg; *Chelonia,* the succulent Green Turtles, so called from the color of their fat; *Eretmochelys,* the Hawksbills, the source of the beautiful tortoiseshell; and *Lepidochelys,* the Ridleys, the smallest of the marine turtles, with a maximum shell length of 790 mm. These turtles are tropicopolitan in the warm seas of the world; stragglers are sometimes carried by warm currents as far north as Newfoundland and Scotland.

Green Turtles were once important commercially as a source of food, but heavy predation by man, on the eggs as well as on the adults, has so reduced their numbers that relatively little turtle meat is obtained for export today. Nevertheless, marine turtles remain a major source of food to natives in many parts of the world. The development of the plastics industry all but eliminated the commercial use of tortoiseshell.

Adult Green Turtles are largely herbivorous, browsing on submerged vegetation in shallow, offshore waters. Loggerheads, on the other hand, are carnivorous, and the Hawksbills and Ridleys are omnivorous.

For a species of animal to survive, its reproductive rate must be sufficient to allow enough of the offspring to reach maturity so that the breeding population is maintained. The number of eggs laid by each female gives some clue to the extent of the hazards faced by the developing young.

FIGURE 15-8
The Atlantic Hawksbill, *Eretmochelys imbricata,*
the source of the tortoiseshell of commerce.

Marine turtles lay many more eggs than freshwater and land forms. During the spring and summer months, the turtles gather offshore from the breeding beaches. At night the female leaves the sea and heaves her heavy body laboriously across the sand to above the high tide mark. Here she digs a nest and deposits her eggs. Two hundred eggs have been found in *Chelonia* and *Eretmochelys* nests, and even the smaller *Lepidochelys* may lay as many as 135. Also, the females apparently deposit more than one clutch during a breeding season—seven emergences for egg laying in two weeks have been reported for a female *Chelonia,* although probably two or three clutches is more usual for marine turtles.

After the eggs are deposited, the female fills the hole, then lumbers back to the safe haven of the sea. The males are waiting offshore, and copulation takes place, so that the eggs for the next breeding season may be fertilized.

The clutches of marine turtles are concentrated in a small area above high tide mark on the laying beaches. Hordes of predators gather to dig them up, the most destructive being man and feral dogs. After hatching, the baby turtles claw upward through the sand and head for the water. On their journey across the beach, they must run the gauntlet of hungry crabs, mammals, and hovering sea gulls. Those that survive are met in the water by swarms of predatory fish. Yet marine turtles have persisted for millions of years (*Caretta* and *Chelonia* are known from the Upper Cretaceous). The marked decline in turtle populations in recent years probably results largely from the activity of human hunters in waylaying the females on the beaches before they deposit their eggs.

SUPERFAMILY DERMOCHELYOIDEA

These marine turtles lack epidermal laminae and have the dermal bones of the carapace and plastron largely replaced by a mosaic of small platelets set in a leathery skin. The limbs are paddle-shaped and without claws: the anterior ones are very large and the posterior ones of the adults are broadly connected to the tail by a web. The skull roof is complete; there is only one biconvex vertebra in the neck. The superfamily includes a single family.

Family Dermochelyidae

A single species, *Dermochelys coriacea,* the Leatherback, is the only living member of the family. This extraordinary creature, one of the most remarkable of all living reptiles, can be confused with nothing else. It is distinguished from all other sea turtles by the scaleless black skin of its back and by the seven narrow ridges, formed by enlarged platelets of the dermal mosaic, that extend down the length of the back. Five similar keels occur on the ventral surface. There is a strongly marked cusp on each side of the upper jaw. This, the largest of the turtles living today, may reach a weight of 680 kg or more, but such giants are rare. The large specimens encountered from time to time along the coasts today probably weigh around 360 kg, about the size of a large loggerhead.

Remarkably strong and rapid swimmers, the pelagic leatherbacks are widely distributed, though usually scarce, in tropical seas, and occur sporadically in temperate waters. They feed on crustaceans, molluscs, and small fishes, as well as on marine plants.

Like the females of other marine turtles, the female leatherback must deposit her eggs on land. She first excavates a broad crater in which her body rests and within that a nest hole. The combined depth of crater and hole may be 90 to 120 cm. After egg deposition, the female fills the hole and crater, then plows a broad area of the surrounding sand thus concealing the nest. Between 80 and 130 eggs are deposited at one time, and there may be two or three clutches during a breeding season.

SUPERFAMILY CARETTOCHELYOIDEA

The epidermal laminae of these turtles are reduced or absent, but the underlying bony shell is complete. The limbs are modified for use as paddles, the digits are long and connected by webbing, and there are only two claws on each foot. The skull roof is emarginate behind, rather than roofed as in the preceding group. There is one biconvex centrum in the neck. Here again this superfamily is represented by only one family.

Family Carettochelyidae

The sole surviving member of the family is *Carettochelys insculpta,* the Pitted Shelled Turtle. Laminae are absent and the pitted shell (about 50 cm long) is covered by only a thin layer of soft skin. This rare and little known freshwater turtle has been found only in the rivers of southern New Guinea. Its life history is unknown.

SUPERFAMILY TRIONYCHOIDEA

The softshelled turtles lack epidermal laminae and their bony shell is somewhat reduced. The shell is low, usually nearly circular in outline, and covered with a leathery skin. The neck is long and retractile, the lips are fleshy (not covered with horny beaks as in other turtles), and the snout is drawn into a fleshy proboscis with the nostrils at the tip. The limbs are paddlelike, with three claws on each foot. The skull is deeply emarginate behind and there are no biconvex centra in the neck. Only one family is known.

FIGURE 15-9
The Hinged Softshell, *Lissemys punctata*, of India.

Family Trionychidae

These extremely aquatic, bottom-dwelling turtles are usually found in rivers, but they may also be found in ponds and swamps. Their length varies from 23 to 80 cm. Most species are carnivorous; they wait for their prey while concealed in mud or sand, or they actively forage. Occasionally plants are

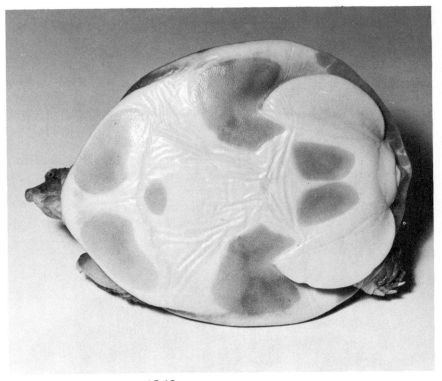

FIGURE 15-10
Ventral view of the Hinged Softshell, showing
pelvic flaps characteristic of this subfamily.

eaten. Softshelled turtles are widely esteemed as food. They are probably also useful as scavengers. Like other turtles with reduced shells, many, though not all, of the species are short-tempered and quick to bite. The living forms are divided into two subfamilies.

Subfamily Trionychinae. The typical softshells have the hyoplastron distinct from the hypoplastron and are without cutaneous femoral flaps on the plastron. There are four genera. The widespread *Trionyx* is found in North America, Africa, and Asia. *Dogania, Chitra,* and *Pelochelys* occur in southeastern Asia, with the last extending to New Guinea.

The breeding habits of the Eastern Spiny Softshell, *Trionyx s. spiniferus* of North America, are probably typical for the family. Nesting occurs in June and July. The female digs a flask-shaped hole, 100 to 250 mm deep and 75 to 125 mm in diameter, with a narrow neck. It takes her about 40 minutes to dig the nest. She deposits a few eggs at a time, arranges them with her feet, and rakes some earth down in the hole with them, gently packing it in, then lays more eggs. This is continued until the nest is finally completed. She apparently makes no effort to conceal the nest after she has covered the last of the eggs. The 10 to 25 eggs are 25 to 27 mm in diameter but are not quite spherical. The shell is thick and not very brittle. The incubation period in this subspecies is not known but eggs removed from an adult female of the Florida race *(T. f. ferox)* hatched in 64 days.

Subfamily Lissemyinae. Members of this subfamily have the hyoplastron and hypoplastron fused. A pair of strong, hinged, cutaneous flaps at the rear of the plastron close over the hind limbs when they are withdrawn. Of the three genera in the subfamily, *Lissemys* is found in India and Burma, and *Cyclanorbis* and *Cycloderma* in Africa. Their breeding habits probably do not differ greatly from those of members of the preceding subfamily. Nesting has been observed in the fall and the smallest young from July to September, early in the monsoon season.

READINGS AND REFERENCES

Agassiz, L. *Contributions to the Natural History of the United States.* Vols. 1-4. Boston: Little, Brown, 1857. (A classic early American work.)

Boulenger, G. A. *Catalogue of the Chelonians, Rhynchocephalians, and Crocodiles in the British Museum.* London: British Museum (Natural History), 1889.

Carr, A. F. *Handbook of Turtles.* Ithaca, N.Y.: Comstock, 1952. (Most readable, and really more than a handbook. Part I is an excellent introduction to the general biology of turtles; Part II is a systematic account of the turtles of the United States and Canada.)

Parsons, J. *The Green Turtle and Man.* Gainesville, Fla.: University of Florida Press, 1962.

Pope, C. *Turtles of the United States and Canada.* New York: Alfred A. Knopf, 1939. (A sound, well-written introduction to the study of turtles, particularly those of the United States.

Pritchard, P. *Living Turtles of the World.* New York: Crown T.F.H. Publications), 1967. (The best illustrated turtle book in print.)

Smith, M. A. *Fauna of British India. Reptilia and Amphibia.* Vol. I. *Loricata, Testudines.* London: Taylor and Francis, 1931. (An exceptionally sound faunal report.)

Wermuth, H. and R. Mertens. *Schildkröten, Krokodile, Brückenechsen.* Jena: Gustav Fischer, 1961. (A recent checklist, worldwide in scope.)

16

LIZARDS AND AMPHISBAENIANS

Order Squamata: Epidermal scales present; vent a transverse slit; male copulatory organs paired hemipenes; no gastralia; teeth either pleurodont or acrodont; vertebrae usually procoelous.

This terse, dry description gives no hint of the extraordinary interest the Squamata have held for man, reputedly since Eve first met the snake in the Garden. The order includes about 5700 different species of the animals that are commonly known as lizards and snakes and comprises three rather closely related groups that are classified as suborders of the large order Squamata. One suborder includes the lizards.

SUBORDER LACERTILIA

It is not always easy to determine whether a given animal is a lizard or a snake by looking at the intact specimen. Most lizards have two pairs of legs, but some have lost one or both pairs, although traces of the girdles remain in all. Usually lizards have visible external ear openings and movable eyelids with a nictitating membrane (third eyelid). The two halves of the lower jaw are firmly united at a mandibular symphysis so that the size of the mouth opening is restricted; the tongue is well developed. The caudal vertebrae of many lizards are divided by a transverse septum at which the lizard is able to break off the tail by muscular contractions (autotomy). This capacity is a defense mechanism; the shed tail part twitches vigorously, thereby distracting the predator while the lizard escapes. The lizard then grows a new tail that is lighter in color and usually smaller than the original, and is supported by a cartilaginous rod rather than by vertebrae.

Lizards can be readily distinguished from the snakes on the basis of internal characters: Legless lizards retain traces of a pectoral girdle and sternum, but no snake has either; the last one or two movably attached ribs in snakes are forked, but lizards never have forked ribs. There are also many differences in the soft anatomy: the kidneys lie far forward in the snake, with the right further forward than the left, but the kidneys of the lizard are more posterior, extending beyond the level of the cloaca, and are symmetrically placed.

As a group the lizards have prospered and now number nearly 3000 species. Although most numerous in the tropics, they have successfully invaded all the continents (except Antarctica) and in Europe they are found as far north as the Arctic Circle. Lizards have undergone extensive adaptive radiation and have developed a bewildering variety of habits. Some spend much of their time in water (either fresh or salt), whereas others have become adapted to life in arid deserts.

The desert-dwelling forms usually have depressed bodies that make it easy for them to hide beneath stones and creep into narrow fissures. Many rock-haunting species have developed special pads on their toes that enable them to run swiftly over smooth surfaces. With the advent of man, some of these shifted their quarters to the walls of houses, where an abundance of insects assures a steady food supply. Lizards usually do not have webs between the toes; the lizard with the most strongly webbed feet, *Palmatogecko rangei*, is found in the arid deserts of southwest Africa where the webbing assists it in maneuvering in the shifting sands.

Some of the ground-inhabiting lizards are large and sluggish. The giant Dragon of Komodo *(Varanus komodoensis)* probably grows to 3 meters in length. Others are small and agile; some run rapidly on two legs (much as small dinosaurs did in the Mesozoic), other ground lizards have lost their limbs, and glide through the grass like snakes.

Some lizards have become burrowers; their limbs are not used in digging through the soil, but are held close to the sides and are usually greatly reduced. A few have lost limbs entirely.

Some arboreal lizards have compressed bodies that make them inconspicuous when they bask on the branches of trees. Most spectacular perhaps are the many species of Flying Dragons *(Draco)* of southeast Asia and the East Indies. A Dragon may be recognized immediately by the winglike expansions of skin on either side of the flank, which are supported by a half a dozen of the hindmost ribs. These ribs function in much the same way as the ribs of a parasol, for when the lizard is at rest it presses them against its body so that the skin is folded and is scarcely noticeable. These expansions are simply gliding surfaces and cannot be used for true flight. The lizard rests on a tree in a vertical position and when it is ready to move it leaps into the air, soars down, and lands with an action so rapid that the opening and closing of the "wings" is scarcely perceptible. One was seen to glide smoothly and steadily for a distance of 18 meters.

The lizards are grouped into six infraorders.

INFRAORDER GEKKOTA

The geckos and their allies are mostly either stoutly built, short-tailed, little lizards, largely nocturnal and well adapted for climbing, or snakelike forms with very reduced limbs. Most geckos are less than 150 mm long. They lack an upper temporal arch, a postorbital arch, and lachrymal, squamosal, and postorbital bones in the skull. The jugal bone is small and sometimes absent. The frontal bones are usually united and surround the forebrain. Pleurodont teeth are usually present on the marginal bones (premaxillaries, maxillaries, and dentaries), but palatal teeth are absent. The eyes have either eyelids or "spectacles," which are simply eyelids that have fused and developed transparent areas through which the lizard sees. The tongue is thick, fleshy, and only very slightly cleft, if it is cleft at all. Most species have a postanal sac opening to the outside on each side of the base of the tail just behind the vent. In the male, a bone lies free just below the skin in front of the opening of the sac on each side. The function of these sacs is unknown. There are usually four or more transverse rows of belly scales per body segment. This infraorder includes those lizards that have well-developed voices.

The Gekkota includes three families. The largest of them (Gekkonidae) comprises those lizards with well-developed limbs that are commonly known as geckos. In one family, the scale-foots (Pygopodidae), the hind legs are represented by scarcely noticeable flaps on either side of the vent and the front legs are lacking. The Xantusidae are known as night lizards.

FIGURE 16-1   *(See facing page)*
Tongues of Lacertilia: (*1*) *Mabuya carinata* (Scincidae); (*2*) *Varanus monitor* (Varanidae); (*3*) *Tachydromus sexlineatus* (Lacertidae); (*4*) *Ophisaurus harti* (Anguinidae); (*5*) *Calotes versicolor* (Agamidae); (*6*) *Gekko gecko* (Gekkonidae); (*7*) *Nessia monodactyla* (Scincidae); (*8*) *Dibamus novae-guineae* (Dibamidae). [After M. A. Smith, 1931.]

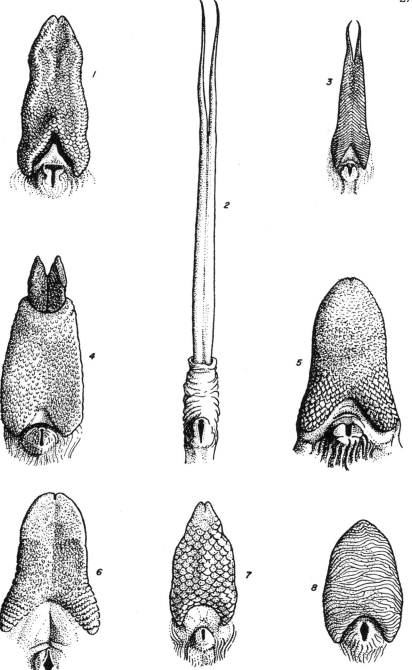

Family Gekkonidae

The geckos form a large tropicopolitan family of about 75 genera and hundreds of species. Most are stoutly built little lizards, nocturnal and arboreal, with big, spectacled, catlike eyes, the pupils contracting to slits in daylight and opening wide at night. The vertebrae are procoelous or amphicoelous. The digits frequently have both claws and friction pads for climbing. The males usually have preanal and femoral pores, as well as postanal sacs and bones.

Many geckos have attached themselves to the dwellings of men. This is really a type of mutualism, since the insects that are also attracted to houses provide food for the geckos and the geckos, by feeding on the insects, help keep these pests under control. Man has been slow to recognize his debt. Indeed, superstitions about them are as widespread as the geckos. It is said that they are highly venomous, that their bite is fatal, that they poison man's food and drink, that they can cause leprosy by running over the face of a sleeper. As with the rat and mouse, man has unwittingly transported geckos about the globe so that the original distribution of many of the genera may never be known.

Few lizards can compete with the geckos in their ability to discard the tail. This defense mechanism is so often used by some species that it is nearly impossible to find an adult specimen with its original tail.

The name "gecko" probably arose as an attempt to imitate the call of some species of these lizards. Their ability to vocalize is remarkable, for most lizards are silent creatures. The sound is perhaps produced by clicking the broad tongue against the roof of the mouth. It has been variously transcribed as "checko," "tocktoo," "toki," "tok," or "chick chick."

Perhaps most spectacular to people seeing geckos for the first time is their ability to run over a window pane or up a vertical wall and across the ceiling. The geckos that are capable of this sort of climbing have some part of their digits dilated to form adhesive discs. In the most arboreal forms the underside of the disc is made of a transverse or fan-shaped series of narrow plates bearing minute, hairlike processes or papillae, which can be pressed into tiny irregularities of the surface. Because the pad does not function well on a smooth surface, a convenient way to collect geckos is with a water pistol. If one squirts the wall on which a gecko is climbing, it will often fall alive and uninjured to the ground.

Most geckos are gentle, but not the Tokay Gecko of Indonesia and the Philippines. When annoyed, the Tokay inflates its body and hisses and puffs loudly, holding its jaws wide open in readiness to attack. If the provocation continues, the lizard rushes forward and seizes some part of its annoyer in its powerful jaws, hanging on with bulldog tenacity. The Tokay is the

FIGURE 16-2
The underside of the Moorish Gecko, *Tarentola mauretanica*,
seen as it clings to a sheet of glass.

largest of the geckos; its head and body are about 175 mm long and its total length is over 300 mm.

Two New Zealand genera in the subfamily Diplodactylinae gave birth to living young but most geckos lay two eggs. These may be hard and somewhat brittle as in the Tokay Gecko or they may be soft and leathery as in *Eublepharis* and *Coleonyx*. It is said that *Tropicolotes algericus* lays but a single egg, as do all of the sphaerodactylines.

Subfamily Eublepharinae.   The Ground Geckos differ from all other geckos except the diplodactylines in that they have true eyelids that can be opened and closed. The digits are straight, and are without friction pads. Postanal sacs and bones are present and the male usually has preanal or femoral pores or both. The vertebrae are procoelous. The tail is short, usually swollen, and somewhat carrot-shaped.

The geographic distribution of members of this family is peculiar. They are widely scattered in desert regions throughout the world. The Banded Gecko, *Coleonyx*, is found from the southwestern United States to Central America; *Holodactylus* is in Somaliland in East Africa; *Hemitheconyx* in Somaliland and in West Africa; *Eublepharis* in Southwest Asia and on islands off the east coast of Asia; and *Aeluroscalabates* in Malaya, Sumatra, and Borneo.

Like the true geckos, and unlike most other lizards, the ground geckos are nocturnal, hiding by day under rocks or in burrows in the sand, coming forth at night to forage for insects. Some have loud voices. Little is known of their breeding, but apparently, like' most other geckos, the female lays two eggs at a time.

Subfamily Diplodactylinae.   The members of this subfamily differ from those in the preceding one primarily in having amphicoelous rather than procoelus vertebrae. This subfamily, with thirteen genera recognized at the present time, is confined to Australia, New Zealand, New Caledonia, and the Loyalty Islands. As mentioned above some members of this subfamily bear living young.

Subfamily Gekkoninae.   The true geckos comprised the largest of the subfamilies in this family with 51 genera recognized at the present time. They lack true eyelids but have a transparent spectacle. The vertebrae are generally amphicoelous and the splenial bone in the jaw is usually present. The voice is well developed, a large Tokay being audible on a quiet night for distances up to 300 feet. They are tropicopolitan in distribution, occurring on all major land masses that reach the tropics and almost all oceanic islands. The female deposits two eggs at a single laying.

Subfamily Sphaerodactylinae.   This is a compact little subfamily of geckos with procoelous vertebrae. The eye is covered by a spectacle. Simple pads are present on the toes of one genus *(Sphaerodactylus)*; the others lack pads. Toes may be straight or angularly bent. Postanal sacs and bones as well as preanal and femoral pores are absent. This subfamily includes the smallest lizards, some of the *Sphaerodactylus* being less than 50 mm long.

The Sphaerodactylinae are restricted in distribution to the tropics of the New World. They tend to be more active during the day than most geckos, and are also more terrestrial. In contrast to other geckos, they are apparently voiceless, and the female, instead of laying two eggs, lays but one hard-shelled egg at a time.

Family Pygopodidae

At first glance, the snake-lizards bear little resemblance to the geckos, but they are very similar to them in many structural characters so it seems best to include them in the infraorder Gekkota. They are slender and snakelike, have no front legs and only a pair of small, scaly flaps near the vent for hind legs. Vestiges of the pectoral girdle remain, and from one to four digits can be recognized in the hind limb. The vertebrae are procoelous, postanal sacs and bones are present, the eye is covered with a spectacle, and the pupil is vertical. The animals are between 150 and 750 mm long. The tail, which is considerably longer than the head and body, is very easily shed.

There are about 8 genera and 14 species of pygopodids, some of them are known from only one or two specimens. They are found only in the Australian region. Little has been learned of their habits. Most are apparently insectivorous but some of the larger species feed on other lizards. So far as is known, all are oviparous.

Family Xantusiidae

The xantusids resemble the geckos in the shape of the vertebrae and in the structure of the eye, which has an elliptical pupil and fused eyelids (the lower eyelid is enlarged into a transparent window). On the other hand, they are like the Scincomorpha in having well-developed temporal arches, with the supratemporal fossa roofed over by the parietal. Postcloacal sacs and bones have been reported for one species but others apparently lack them.

The family contains only four genera; *Xantusia* is found in southwestern United States, *Klauberina* on the Channel Island group off the coast of southern California, *Lepidophyma* in central Mexico and Central America, and *Cricosaura* in Cuba.

These are little lizards with well-developed legs and tails. The dorsal scales are granular, the belly scales rectangular plates. The tail is easily shed. Secretive and nocturnal, they are most often found, in the United States at least, hiding under the flakes of rock that form on the faces of granite boulders in desert regions. They are probably insectivorous.

Xantusids are truly viviparous. Mating of *Xantusia vigilis*, the Desert Night Lizard, takes place in May, and ovulation occurs one to four weeks later. Usually two eggs are formed at one time (this is another way in which the Night Lizards resemble the geckos), but rarely only one embryo is found (it is usually in the right oviduct). Occasionally, however, three embryos are present. Early in embryonic development, a simple, cellophanelike shell is formed, but it soon disintegrates, and a chorioallantoic placenta develops. Gestation takes about three months. Surprisingly enough, these lizards have developed the typically mammalian custom of eating the fetal membranes. The membrane ruptures before the young is born and remains in the cloaca. The female grasps the protruding edge in her mouth, gradually draws it out, and swallows it.

Two subfamilies are recognized.

Subfamily Cricosaurinae.   These have two frontonasal scales, a single frontal, no parietal scales; the fourth finger has four phalanges. Only *Cricosaura* of Cuba belongs here.

Subfamily Xantusiinae.   Members of the three other genera of night lizards have one frontonasal, two frontal, and two parietal scales. The fourth finger of the hand has five phalanges.

INFRAORDER IGUANIA

This infraorder, and the Scincomorpha, include the majority of the animals of which most people think when they hear the word "lizard." The iguanians are numerous and varied, some being large, others quite small, and they are frequently brightly colored, and often ornamented with crests, spines, frills, and throat fans. They are diurnal and may be either arboreal or terrestrial. *Amblyrhynchus* of the Galápagos Islands is semimarine. No iguanian shows any tendency toward the development of a snakelike body or a reduction of the limbs.

The temporal arch is present, the skull is high, the teeth are either pleurodont or acrodont, the parietals are fused to form a single bone. There are six cervical vertebrae, and four or more rows of transverse belly scales per body segment. The tongue is simple rather than divided into anterior and posterior portions. The eyelids are well developed, and the pupils are round. Femoral pores are occasionally present. Breaking point septa are present in the caudal vertebrae and some of these lizards have rather fragile tails.

This infraorder contains two very similar families—Iguanidae and Agamidae—which are essentially New and Old World counterparts of each other.

Family Iguanidae

This is the largest family of lizards of the New World. There are two genera *(Chalarodon* and *Oplurus)* in Madagascar and one *(Brachylophus)* in Polynesia, on the Fiji and Tonga Islands, but there are more than 50 genera in the western hemisphere. These lizards have a pleurodont dentition in which the teeth are attached on the inner surface rather than along the dorsal

FIGURE 16-3
The Iguana, *Iguana iguana.*

margin of the lower jaw. The splenial bone of the lower jaw is well developed in contrast to its reduced condition in the sister family, Agamidae.

The iguanids range from less than 125 mm long to the giant iguanas which may be 1800 mm long. These large iguanas are used as food in some tropical countries. The smaller iguanids are mostly insectivorous or carnivorous, but many of the larger ones are herbivorous. The marine iguanas of the Galápagos enter salt water to forage for seaweed, which seems to be their favorite food.

A characteristic habit of the family is head-bobbing which, in such forms as *Anolis*, is correlated with the distention of a throat fan. The head is raised above the surface and bobbed backwards so that the brightly colored fan is spread, but the characteristic bobbing motion occurs also in iguanids that do not have a throat fan.

Iguanids are usually oviparous, but some species of the Horned Lizards *(Phyrnosoma)* and some of the Spiny Lizards *(Sceloporus)* are ovoviviparous. The eggs are soft-shelled and are frequently buried in the ground, even by species that are largely arboreal.

FIGURE 16-4
The Cuban Anole, *Anolis sagrei*. The bifid tail apparently resulted from an injury.

Family Agamidae

This large family is the Old World counterpart of the New World Iguanidae. They are most numerous in the Oriental Region, but are found also in extreme southeastern Europe, Africa, Australia, and the New Guinea archipelago. They are absent from the only places in the Old World from which iguanids are known: the Fijis, Tongas, and Madagascar. Thus the two families are entirely separate in geographic distribution.

The Agamidae differ from the Iguanidae in the structure of their jaws and teeth. The teeth are acrodont (attached to the upper margin of the jawbone), and the dentition is heterodont (divided into incisorlike, caninelike and molarlike teeth). We say "like" since these are probably not homologous to the similar teeth in mammals. Many species have well-developed ornamental crests, frills, or throat pouches, frequently brilliantly colored. The "wings" of the Flying Dragons are as bright as those of some butterflies. Most agamids are of moderate length, but *Hydrosaurus*, the water lizard of the East Indies and New Guinea, may reach 900 mm, and the Toad-headed Agamids, *Phrynocephalus*, may be less than 125 mm.

Like the iguanids, the agamids have undergone adaptive radiation: some are terrestrial, some arboreal, some verge on fossorial. Most are carnivorous, some are omnivorous, and the rock-dwelling *Agama melanura* and *A. nupta* seem to be herbivorous. One, *Uromastix*, is largely herbivorous as an adult although it often eats insects.

FIGURE 16-5
This Mastigure, *Uromastix acanthinurus werneri*, refused food until offered the zinnia, whereupon it bit off the petals as fast as it could swallow them. It is a resident of northern Africa.

The Agamidae seem to be mostly oviparous but at least two genera, *Phrynocephalus* of Central Asia and *Cophotes* of Ceylon, are ovoviviparous. The Flying Dragons, although otherwise entirely arboreal, descend to earth to bury their eggs.

INFRAORDER RHIPTOGLOSSA

This infraorder contains the most weird-looking of all lizards, the true chameleons. They are highly modified for arboreal life, with short, compressed bodies, and coiled, prehensile tails. The skull may be ornamented with horns, crests, and tubercles, and produced backward to form a grotesque, helmetlike casque. Most chameleons are moderate in size; various species range from less than 150 to 600 mm in total length.

The dentition is acrodont, there is a dorsal temporal arch, and the parietals are fused to form a single bone. There are only three cervical vertebrae. The tongue is elongated and enlarged at its distal end, but it is not divided into distinct fore and aft sections. There are four or more rows of transverse scales on the ventral side of the body for each body segment. The tail lacks breaking points. The thick, granular eyelids are united except for a small slit in the center. The hands and feet are zygodactylous (with yoked digits). The two inner fingers of the hand are bound together in a bundle which opposes the similarly joined three outer fingers. The three inner toes of the feet oppose the two outer toes. This forms a very efficient grasping mechanism.

Family Chamaeleonidae

This is the only family in the infraorder. It is divided into six genera; more than 80 species are recognized. The vast majority live in Africa and Madagascar, but the Common Chameleon of northern Africa *(Chamaeleo chamaeleon)* is also found on some of the islands of the Mediterranean and ranges northward into southern Spain and Portugal. Two species occur in southwestern Asia and one inhabits the Indian region.

A chameleon is able to move each of its round, protuberant eyes independently. This ability is slightly developed in some iguanians, but only in the true chameleons has it reached the point where one eye is entirely independent of the other. It is fascinating to watch a chameleon resting on a limb, with the left eye rolling in one direction while the right eye rolls around peering in the other. When a large insect or small bird is seen, both eyes focus on it, and the lizard creeps along with painful slowness until it is within a few inches of its prey. Then, faster than the human eye can follow, the long, extensile tongue shoots out, attaches to the creature, pulls it into the mouth, and with a gulp the victim is gone.

Chameleons are famous for their ability to change color, from white through shades of yellow, green, and brown, to black, variously spotted and blotched with contrasting colors. Some iguanians, such as *Anolis*, show a similar ability to change color, and it is because of this the anoles are often called chameleons. These color changes are responses to changes in light, heat, and the emotional state of the animal, and not, as is popularly believed, to the color of the background.

Some chameleons are oviparous, others ovoviviparous. A pair of captive *Chamaeleo zeylanicus* from India were mated during the first week in October. The female would not allow the male to approach her after mating had occurred. A little more than a month after copulation, she descended to the ground and began digging like a terrier, packing the loose earth with her forelegs and kicking back with her hind legs. She remained in the hole overnight and continued to work on the nest the following day. About 2:00 o'clock in the afternoon of the third day she emerged and spent the rest of the day pulling the earth back with her forelegs and packing it tightly with her hind legs. She did not finish filling the nest until the morning of the fourth day. The 31 oval, soft-shelled eggs measured 13 by 7 mm; they were buried about 30 cm below the surface of the ground. Eggs as large as 19 by 12 mm have been found in dissected specimens.

INFRAORDER SCINCOMORPHA

This is a large, cosmopolitan group of lizards, medium to small in size, with a strong tendency toward reduction of the limbs and development of a snake-like body habitus. The degenerate forms are usually burrowers. In the Scincomorpha, the dentition is pleurodont, the dorsal temporal arch is usually present, and the parietals are fused. The tongue is simple. There are six cervical vertebrae and fewer than four rows of transverse belly scales per body segment.

The infraorder is cosmopolitan in distribution and comprises seven families.

Family Scincidae

The abundant and ubiquitous true skinks are usually small, secretive, semi-burrowers with highly polished scales. Most are less than 200 mm long, and the largest, *Corucia zebrata* of the Solomon Islands, is only about 600 mm in length. The temporal openings of the skull are more or less roofed by backward growths of the postfrontals. Pterygoid teeth are often present. The limbs may be present or absent; those species that have lost their limbs always possess some traces of the pectoral and pelvic girdles. Abdominal and parasternal ribs are sometimes present, chiefly among the burrowing forms. The head, body, limbs, and tail are protected by osteoderms. The head is covered with symmetrical shields, the pupil is round, femoral pores are absent. The tail is easily broken but another is quickly regenerated. Some of the skinks have developed a transparent window in the lower eyelid which enables the lizard to see while the eyelids are shut. The most extreme development along this line is found in the little, active, snake-eyed skinks *(Ablepharus)* whose eyelids are wholly transparent and immovable, permanently covering the eyes like a pair of watch glasses. Not all skinks are smooth and shiny; a few bizarre-looking forms have big, spiny scales on their backs, sides, and tails. The stout-bodied, stubby-tailed Shingle-back of Australia *(Trachydosaurus)* looks like an animated pine cone.

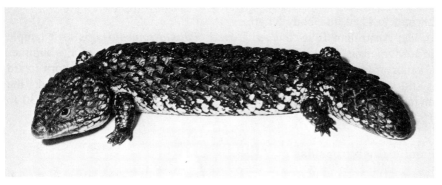

FIGURE 16-6
The bizarre Shingle-back of Australia, *Trachydosaurus rugosus.* The short tail is responsible for the common name "stump tail," often applied to these skinks.

Although they are cosmopolitan in distribution, skinks are most numerous in Australia, the islands of the western Pacific, the Oriental Region, and Africa, and are poorly represented in the Americas. There are nearly 50 genera, and more than 600 species.

The vast majority of skinks are terrestrial and are usually extremely active; a few have arboreal tendencies, but none is so highly adapted for climbing as are members of the preceding infraorders. Some skinks have become rock dwellers and a few are good swimmers. A few, like *Neoseps* and *Ophiomorus,* have become specialized for living in loose sand and move through it with a swimming motion. Skinks are often found under piles of dead leaves, coconut husks, and rotting vegetation or decaying logs. The terrestrial skinks are diurnal whereas the burrowers seem to be largely crepuscular or nocturnal. Most skinks are insectivorous, but some of the larger ones also consume small vertebrates, and a few are partly herbivorous.

Skinks may be either oviparous or ovoviviparous, and a few, such as the European skink, *Chalcides ocellatus,* approach true viviparity by having a placenta. *Eumeces fasciatus* of the eastern United States is a typical oviparous skink. In Maryland, courting and copulation take place shortly after emergence from hiberation, generally during early May; the eggs are laid 6 or 7 weeks afterward. The clutches range from 2 to 18 in number, the smaller clutches being laid by the smaller females. There is also an indication that the size of the clutch decreases toward the northern part of the range. The eggs are deposited in rotten wood or loose soil 50 to 75 mm below the surface and are brooded for the entire incubation period by the mother. Shortly after deposition, the eggs are approximately 13 by 7 mm but, like the eggs of most lizards (except the geckos), they increase in size during incubation and shortly before hatching may be as much as 20 mm long. They hatch 4 to 7 weeks after deposition. At hatching the young are 24 to 28 mm in head and body length.

The Australian Blue-tongued Lizard, *Tiliqua scincoides*, is an example of a viviparous skink. The young are born in the middle of the summer (January); litters of from 5 to 18 have been recorded. The newborn lizard is still wrapped in the fetal membrane, but in a few seconds it breaks the membrane and immediately devours it. At birth the young are from 130 to 152 mm in total length.

Family Anelytropsidae

The only genus in this family is a small (200 mm) wormlike, limbless lizard, presumably derived from the Scincidae, but differing in the absence of a pectoral girdle and the loss of the temporal arch from the skull. The single species, *Anelytropsis papillosus*, occurs in central and eastern Mexico. Only three or four specimens have been collected; one from under a rotten log near an ant nest and another from under a rock. Nothing is known of its way of life.

Family Dibamidae

The family comprises a single genus of slender, wormlike lizards not more than 225 mm long. Only vestiges of the pectoral girdle remain; the hind limbs too have disappeared in the female though they are represented by scaly flaps in the male. The eye is greatly reduced and covered by skin. The tail is short and apparently cannot be shed. The egg has a calcareous shell. *Dibamus* ranges from southern Indochina and the Philippines to the New Guinea archipelago.

Family Feylinidae

This family was erected for a genus of small, wormlike, limbless scinco-morphs. Like *Anelytropsis*, members of the genus lack skull arches, but they retain a vestige of the pectoral girdle and are somewhat larger, being about 300 mm long. Chevron bones are associated with the caudal vertebrae. The four species of *Feylinia* are found only in equatorial Africa. They feed almost exclusively on termites and are most often found under rotten logs where these insects abound. Their life history is unknown.

Family Cordylidae

This small family of scincomorph lizards is somewhat intermediate between the Scincidae and the Lacertidae, the next family to be described. It includes genera that were formerly classified in two separate families—Cordylidae (or Zonuridae) and Gerrhosauridae. These genera are quite similar, both in structure and in geographic distribution, and it seems best to put them together in a single family. These again are small to medium-sized lizards, the largest of which is only a little more than 600 mm long, with the tail comprising most of the length. Osteoderms are present on the head and body, the limbs are sometimes reduced to stumps, and femoral pores are well developed. The tongue has papillae, but is only feebly nicked anteriorly. The ten genera are resticted to Africa south of the Sahara and to Madagascar. Both oviparous and ovoviviparous cordylids are known. *Gerrhosaurus v. validus* of southern Africa lays about four soft-shelled eggs in September or October, whereas *Cordylus cordylus* of the same region bears one or two living young.

Family Lacertidae

The members of this family are usually small, agile, long-tailed lizards. Alone among the Scincomorpha, they show no tendency toward reduction of the limbs. The dorsal temporal arch is complete, but bony dermal plates (osteoderms) cover the temporal fenestra and fuse with the cranial bones when in contact with them, thus obscuring the structure of the dorsal arch. Osteoderms are lacking on the body, though they are present on the head. The lateral teeth are often bicuspid or tricuspid. The tongue is moderately elongated, deeply notched anteriorly, and covered with scalelike papillae or with transverse plicae. Femoral glands are usually present. Some forms have windows in the lower eyelids. The largest species is the Jeweled Lacerta *(Lacerta lepida)* which reaches a length of 750 mm.

The Lacertidae are Old World Lizards, occurring in Europe, Asia, and Africa, but not in Madagascar or in the Australian Region. They are most abundant in Africa and comparatively rare in the Oriental Region. The Common Lizard of Europe, *Lacerta vivipara,* exists above the Arctic Circle; no other lizard is found this far north. The family includes about 20 genera.

Lacertids are predominantly terrestrial, often living in grassy or sandy places. Some, such as the European Wall Lizard *(Lacerta muralis)*, are agile climbers. They are carnivorous, feeding chiefly on insects and other small invertebrates.

Lacerta vivipara, as its name indicates, usually bears living young, but the other lacertids are oviparous. The Sand Lizard, *Lacerta agilis agilis*, breeds during May and early June; the same male and female may mate many times during this period. The eggs, which are laid in June and July, vary from 6 to 13 in number, depending in part on the size of the female. When first laid, the eggs are 12 to 15 mm long and 8 to 9 mm wide; shortly before hatching they are 15 to 20 mm long. The eggs are hidden under stones or in shallow holes dug in the earth and covered over by the mother. They hatch in 7 to 12 weeks and the young, at hatching, are 56 to 63 mm in head and body length.

Parthenogenesis in lizards was first reported in the Rock Lizard, *Lacerta saxicola* in Armenia. It has since been recorded in a number of all female or virtually all female species of the genus *Cnemidophorus* in the American southwest. *Cnemidophorus* is a member of the family Teiidae, a New World counterpart of the Old World lacertids.

Family Teiidae

This family occupies in the New World a place similar to that filled in the Old World by the closely related Lacertidae. The largest of the teiids *(Tupinambis)* is about 900 mm long, but most are small. Some are degenerate burrowers with reduced limbs. Osteoderms are absent from the head and the body. The skull arches are present and the temporal fenestra are open. The tail is quite long. The tongue, like that of the Lacertidae, is long and narrow, deeply forked anteriorly, and covered with papillae. The front teeth are always conical but the lateral teeth on both jaws may be conical, bicuspid, tricuspid, molariform, or even enormous, oval crushers (as in the snail-eating Caiman Lizard, *Dracaena guianensis*).

The family is found only in the western hemisphere, where it is represented by 40 genera and about 200 species. Practically all are restricted to South America; only one, *Cnemidophorus*, reaches the United States.

Teiids live in a wide variety of habitats, from tropical jungles to deserts and from the seashore to the high Andes. Most are either terrestrial or fossorial but *Dracaena* is semiaquatic. The larger species are diurnal, but some of the smaller ones are nocturnal. Most, but not all, are carnivorous.

Life history notes are fragmentary, but observations made on several different species seem to indicate that on the whole the teiids have rather uniform life histories. A female of the Coastal Whiptail, *Cnemidophorus tigris multiscutatus*, is known to have mated on May 23 and again on May 25. About 3 weeks later, on June 13, she laid 3 eggs. She mated again on July 3 and on July 22 she again laid 3 eggs. They were immaculate white and were 19.5 to 20.8 mm in length and 10.9 to 11.0 mm in width. They

FIGURE 16-7
The snail-eating Caiman Lizard, *Dracaena guianensis*.

increased about 8 percent in length and 35 percent in width during the incubation period of about 80 days. The young were between 110 and 118 mm long at hatching.

As noted above some *Cnemidophorus* are parthenogenetic.

INFRAORDER ANGUINOMORPHA

This infraorder includes a heterogeneous assemblage of animals ranging from the bulky, ten-foot-long Komodo Dragon, *Varanus komodoensis*, to the wormlike California Legless Lizard, *Anniella pulchra*, no bigger than a lead pencil. At first glance, they seem to have nothing in common, but underlying structural similarities justify their being placed together. The tongue is divided into two parts, with a notched, inelastic forepart separated by a transverse fold from the elastic hind part, which serves as a sheath when the tongue is withdrawn. The teeth are nearly solid, not hollowed at the base as are those of other lizards, and are replaced alternately (that is, the new tooth comes up behind, not beneath the older tooth). There are relatively few anguinomorphs living today, but the group includes a number of large, heavily armored, extinct forms, notably the huge, aquatic mosasaurs, the most spectacular lizards that ever lived.

The infraorder is divided into two superfamilies, each having three living families.

SUPERFAMILY DIPLOGLOSSA

These are medium-sized to small insect-eating lizards, generally more or less armored with osteoderms. Many have reduced limbs and snakelike bodies. The external naris is not prolonged backward as a slit but is a round or oval foramen. The toothed maxillary extends far back beneath the orbit and the bones of the lower jaw are rigidly joined together. The tail can be regenerated.

Family Anguinidae

This family includes some lizards with well-developed limbs, such as the Alligator Lizards *(Gerrhonotous)*, and some that lack limbs, such as the Glass Lizards *(Ophisaurus)* and Slow Worms *(Anguis)*. The tropical American genera known as Galliwasps show intermediate stages in limb reduction. The temporal arches are present, the temporal openings are long and narrow and in some forms are roofed. Palatal teeth may be present or absent. Osteoderms are well developed. Most anguinids are medium sized; the largest, *Ophisaurus apodus*, may attain lengths of nearly 1200 mm.

This family includes about ten genera, most of them found in tropical America. Only two are present in the Old World: *Ophisaurus* occurs in Asia, southeastern Europe, and northern Africa, as well as North America; *Anguis* is found only in Europe, northern Africa, and western Asia.

Most of the anguinids are terrestrial, a few are somewhat arboreal. Some of the species hide in burrows during the day, and go forth at night to hunt for insects and other small invertebrates.

Both oviparous and ovoviviparous forms occur, not only within the family, but within a single genus. The San Francisco Alligator Lizard, *Gerrhonotus coeruleus coeruleus*, which is ovoviviparous, copulates in April. The copulatory process is lengthy, sometimes lasting for many hours. The number of young varies from 2 to 15, but usually about 7 are born in late August or September. When first born, they have a snout-to-vent length of 25 to 30 mm. On the other hand, the Oregon Alligator Lizard, *Gerrhonotus multicarinatus scincicauda,* is oviparous. Mating occurs from the middle of May to the middle of June. The eggs, which may number more than a dozen in a single clutch, are laid in late July and early August, in burrows dug by mammals. The young hatch in September.

Family Anniellidae

This family is represented solely by *Anniella*, the little, shovel-snouted, Legless Lizard of California and Baja California. The bones of the skull are closely knit, there are no temporal arches, the snout is short, and the brain-case is expanded. There are no palatal teeth and the osteoderms are reduced. This lizard is about 250 mm long, lacks external ear openings, but has lidded eyes. The body is covered on all sides with smooth, rounded, uniform, over-lapping scales like those of a skink.

Legless lizards lead almost exclusively underground lives but are some-times found on the surface of the sand under rocks and logs.

Anniella is ovoviviparous; one to four young are born in late summer and fall.

Family Xenosauridae

Two poorly known genera, widely separated geographically and very limited in distribution, are included in this family. *Xenosaurus* is found from southern Mexico to Guatemala; *Shinisaurus* occurs in southern China. The two genera are sometimes placed in different subfamilies, Xenosaurinae and Shinisaurinae. They are medium sized (about 250 to 375 mm), with nor-mally developed limbs. Two clearly defined, longitudinal crests, formed by series of enlarged scales, run down the midline of the back. The rest of the dorsal surface is covered by a scattering of smaller scales that are inter-spersed with minute granules. The temporal arches are strongly developed and the temporal openings are large and not roofed by skull bones. The bones of the skull are roughened by the fusion to them of the cranial osteoderms.

Shinisaurus lives along streams and feeds partly on tadpoles and fish. It also basks on tree branches overhanging water.

SUPERFAMILY PLATYNOTA

This superfamily includes medium-sized to very large predaceous lizards, with jaws adapted more for grasping large prey than for chewing small invertebrates. Thus, the bones of the mandible are rather loosely joined and there is a tendency toward the development of a hinge within the jaw. The maxillary bone barely reaches the level of the orbit, and the marginal teeth are thus in front of the eye. The slitlike external nares extend far back in the skull. The tail is not autotomous and the limbs are never reduced. Three families are included in the superfamily, each has only one genus.

Family Helodermatidae

The Gila Monster *(Heloderma suspectum)* and the Mexican Beaded Lizard *(Heloderma horridum)*, found only in southwestern United States and

Mexico, are the only venomous lizards known. Unlike those of the poisonous snakes, the venom glands of *Heloderma* are in the lower jaw and are not connected with the teeth. The venom empties into the mouth through several ducts that open between the teeth and the lips. Grooves on the teeth help draw the venom into the wound by capillary action as the lizard hangs on and chews. The bite is painful, but seldom fatal to man. There are a few palatine and pterygoid teeth, no temporal arches, and eight cervical vertebrae. The back and the outer surfaces of the limbs are covered with large osteoderms.

FIGURE 16-8
The Gila Monster, *Heloderma suspectum*. It and its relative,
H. horridum, are the world's only venomous lizards.

These are heavy-bodied, short-tailed, clumsy looking lizards, gaudily marked with dark reticulations on a yellow or orange background or vice versa. The Beaded Lizards may reach a length of 900 mm. The Gila Monster is shorter, being at most about 500 mm long. In captivity both feed entirely on hens' eggs so it seems probable that eggs and nestling birds form a major part of their natural diet. Poison seems to be unnecessary against such prey, and the function of the venom remains obscure.

Heloderma suspectum has been seen in copulation in mid-July. What little evidence there is indicates that three to seven eggs are laid during a period from late July to mid-August. They are buried to a depth of about 125 mm in an open place that is exposed to the sun but usually near a stream or dry wash. The thin-shelled, white, rather rough eggs are 67 to 75 mm long and 33 to 39 mm wide. The incubation period is apparently 28 to 30 days.

Family Varanidae

All living members of this family are included in a single genus, *Varanus*, the monitor lizards. The smallest monitor, *Varanus brevicauda* of Australia, is only about 20 cm long, but most are large, so much so that they have been mistaken for crocodiles. Indeed, the native name for the Komodo Dragon, "buaja darat," means land crocodile. This monitor may reach a length of 300 cm. Monitors lack the venom apparatus of the helodermatids. The dorsal temporal arch is complete, there are no palatine and pterygoid teeth, and the osteoderms are reduced or sometimes lacking. The tail is long and muscular and, like that of the crocodile, is a formidable weapon. There are nine cervical vertebrae. A monitor holding its head erect on its long neck has a very alert appearance.

About 30 species of *Varanus* are known to occur in southern Asia, Africa, the East Indies, and Australia. All are carnivorous; the larger forms have been reported capable of tackling such prey as pigs and small deer. Although largely terrestrial, they are surprisingly agile climbers for such bulky animals. Many are quite aquatic and have been seen swimming far at sea. This undoubtedly explains their distribution throughout the East Indies.

The monitors are oviparous. In Thailand, the Common Water Monitor, *Varanus salvator*, lays 15 to 31 eggs at the beginning of the rainy season in June. They are deposited in holes in riverbanks or perhaps in trees beside the water. They measure about 70 by 40 mm, have rather soft shells, and are said to taste like turtle eggs.

Family Lanthanotidae

The sole representative of this family, *Lanthanotus borneensis*, is known only from the vicinity of a few streams in Borneo. One of the rarest of lizards, at least in museum collections, it is a dull-colored, short-legged animal about 400 mm long. The back and tail are protected by ridges of raised tubercles. The head is covered with small, grainlike nodules. There is no dorsal temporal arch, there are a few teeth on the palatine and pterygoid bones, and there are nine cervical vertebrae. These animals bear a superficial resemblance to *Heloderma* and were long classified with that genus, but they have no venom apparatus and are anatomically much closer to *Varanus*. The Earless Monitor is apparently nocturnal and to some extent fossorial. Captive specimens have eaten fish and the eggs of turtles.

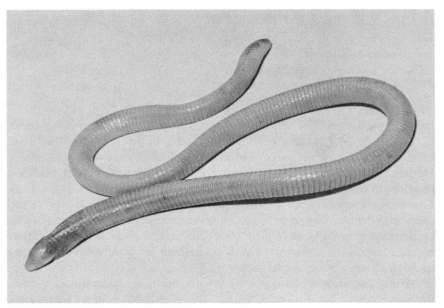

FIGURE 16-9
The Florida Worm Lizard, *Rhineura floridana*, locally known as "Graveyard Snake."

SUBORDER AMPHISBAENIA (ANNULATA)

Included in this suborder are a number of small, highly specialized burrowers that do not seem to be closely related to either lizards or snakes. Their bodies are elongate and nearly uniform in diameter, and their tails are short. Except for one genus, *Bipes,* which has short front legs, external limbs are lacking and the girdles are vestigial. There are no external ear openings and the eyes of the adults are hidden under the skin. Osteoderms are absent. The soft skin of the body is folded into numerous rings divided into quadrangular areas representing the flattened and reduced scales. These rings, combined with the absence of limbs and the cylindrical body form, give the Annulata a remarkable resemblance to earthworms.

The skull of an amphisbaenian is highly specialized for digging: the bones are closely united, the snout region is short and usually expanded, the braincase is almost completely enclosed. There are no skull arches. The teeth are large but few in number, and are absent from the palate. There is only one functional lung, the left. Most of the species are about 300 mm long. The largest, *Monopeltis* of Central Africa, attains lengths of about 675 mm. There are two families.

Family Trogonophidae

This is a small family with four genera and about six species. They are restricted to northern Africa and Socotra Island in the Indian Ocean. Their life histories are not well known but *Trogonophis* gives birth to living young rather than laying eggs as do the members of the other family, Amphisbaenidae.

Family Amphisbaenidae

This family contains 19 genera and about 35 species. Members are found in Africa and the Mediterranean countries, in South America and the West Indies, and range north through Mexico to Baja California and possibly to Arizona. One species, *Rhineura floridana,* is found in Florida.

The name of the family comes from two Greek words which, when translated literally, mean "walk at both ends," a reference to the ability of amphisbaenids to move backward as well as forward in their underground tunnels. A most appropriate common name is Worm Lizard. In Florida, where they are best known to grave diggers who uncover them while digging in the sandy soil of cemeteries, they are sometimes called Graveyard Snakes. They feed on worms and small insects, especially ants and termites. So far as is known, most amphisbanids lay eggs though one form has been reported to be ovoviviparous.

READINGS AND REFERENCES

Boulenger, G. A. *Catalogue of the Lizards in the British Museum (Natural History).* London: British Museum (Natural History), 1885–1887.

Camp, C. L. "Classification of the Lizards." *Bulletin of the American Museum of Natural History,* vol. 48, art. 11, 1923. (A sound, basic work in lizard classification.)

Cope, E. D. *The Crocodilians, Lizards and Snakes of North America.* Report of the United States National Museum for 1898. Washington: Smithsonian Institution, 1900. (A basic work on North American forms.)

Gans, C. "A Check List of Recent Amphisbaenians." *Bulletin of the American Museum of Natural History,* vol. 135, art. 2, 1967.

Kluge, A. "Higher Taxonomic Categories of Gekkonid Lizards and their Evolution." *Bulletin of the American Museum of Natural History,* vol. 135, art. 1, 1967.

McDowell, S. B., Jr., and C. M. Bogert. "The systematic position of *Lantho-notus* and the affinities of the Anguinomorphan lizards." *Bulletin of the American Museum of Natural History,* vol. 105, art. 1, 1954. (A modern treatment of one infraorder.)

Smith, H. M. *Handbook of Lizards.* Ithaca, N.Y.: Comstock, 1946. (A modern handbook, restricted to North American forms.)

Smith, M. A. *Fauna of British India. Reptilia and Amphibia.* Vol. II. *Sauria.* London: Taylor and Francis, 1935.

Underwood, G. "On the classification and evolution of geckos." *Proceedings of the Zoological Society of London,* vol. 124, part 3, 1954.

17

SNAKES

An interest in snakes has probably led more people to study herpetology than an interest in all the other groups of herptiles combined. Although it is true that many of these students have later turned to more prosaic forms, such as the salamanders or turtles, it is still the snakes that first attracted their attention.

The approximately 2700 kinds of snakes in the world form the suborder Serpentes of the order Squamata. Like the lizards in the suborder Lacertilia, they have undergone extensive adaptive radiation and have come to occupy most of the major habitats of the world. Some, the burrowers, are usually small and have eyes that are hidden beneath the scales of the head. Others have taken to the trees and seldom come to the ground. One group is entirely marine; these include the only modern reptiles to abandon the land completely (the sea turtles come to shore to lay their eggs). In size the snakes range from tiny forms only about 100 mm long to the Anaconda and the Reticulated Python which reach lengths of more than 9 meters.

Snakes are not as important economically as turtles. Some, probably most, are edible; but they are seldom used as food, although in the Orient not only the Pythons but also the poisonous sea snakes are esteemed by some peoples. Snake skins make a leather suitable for fancy belts, pocketbooks, and shoes, but not much else. Structurally, the snakes are so highly modified that they are not popular specimens in anatomical laboratories, and indeed there is no good description of the anatomy of a snake comparable to those available for the salamander and the frog. Their major contribution to mankind is the control of pests, particularly the destructive rats and mice.

Snakes are elongate animals with either no girdles or limbs or occasionally with vestigial pelvic girdles and hind limbs. They lack a sternum, external ear opening, tympanic membrane, middle ear, and eustachian tube. Except in some burrowing forms, the immovably fused and transparent eyelids form a protective window, the brille, beneath which the eye moves. The viscera are elongated, and the left lung is smaller than the right or altogether absent. The tongue is long, forked, and protractile. There is no urinary bladder. Like the lizard, the snake has a transverse vent, paired copulatory organs, and a body that is covered with scales.

The skull of a snake is more specialized than that of a lizard. The brain cavity is completely enclosed anteriorly by dermal bones. Higher forms have the bones of the facial region and jaws loosely joined to each other and to the cranium so that they can be spread apart. The two halves of the lower jaw are not fused but are connected by a ligament. Each half of both upper and lower jaw can be moved independently of the other half. This allows a snake to engulf objects that look impossibly large. It seems to "walk" its

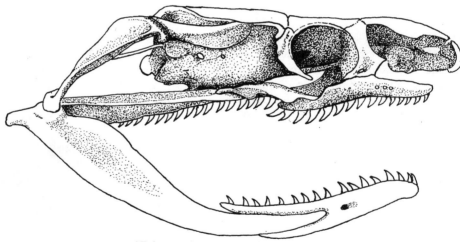

FIGURE 17-1
The skull of the colubrid snake *Drymarchon corais couperi*,
the Indigo Snake of the southeastern United States.

mouth around its food by a forward movement of first one side of the mouth and then the other.

The elongation of the vertebral column results from an increase in the number of vertebrae rather than from a lengthening of individual vertebrae.

The earliest reptiles seem to have had about 25 presacral vertebrae. Because of the absence of the girdles, it is difficult to classify the snake vertebral column into regions. The total number of vertebrae reported for different species ranges from 141 to 435. Most of these are dorsals. The caudals (i.e., those located behind the region of the vent) average about 50 to 60 in number, although in the short-tailed burrowing forms the number is much lower and in some long-tailed snakes the ratio of dorsals to caudals may be only 1:1.

The vertebrae have a complicated structure. In addition to the usual articulating facets—the prezygapophyses and postzygapophyses—each has a pair of zygantra and a pair of zygosphenes (see Fig. 3–7). Thus each vertebrae has five points of contact with the anterior one: the centrum, the two prezygapophyses which fit against the postzygapophyses, and, dorsal to the zygapophyses on the neural arch, the zygosphenes which articulate with the zygantra. The vertebrae of the anterior part of the trunk have hypapophyses—ventrally directed processes from the centra—to which muscles are attached. In many groups these are present on the posterior dorsals as well. In place of the hypapophyses, the caudal vertebrae frequently have haemo-pophyses—paired ventral projections from the centra that surround the caudal blood vessels.

The snakes are divided into three infraorders.

INFRAORDER SCOLECOPHIDEA

This infraorder includes a number of small, burrowing snakes that are very primitive in some ways, very specialized in others. The vertebrae lack neural spines. The optic foramen is surrounded by the frontal bone. Usually traces of the pelvic girdle are present though there are no external spurs representing the hind limbs. The liver is composed of a great number of lobes whereas in other snakes it is almost unlobed. There is only one pair of thymus glands; other snakes have two pairs. There are three families in this infraorder.

Family Typhlopidae

These are wormlike burrowing creatures with cylindrical bodies and short tails. The largest species may be 750 mm long but most are less than 200 mm. The eyes are more or less distinct but are covered by head shields. The transversely placed maxilla is loosely attached to the skull and bears teeth that are directed backward. The premaxillary, palatine, and pterygoid bones lack teeth. Some species have a single tooth at the tip of each mandible, but there is never a row of teeth on the lower jaw. Zygosphenes and zygantra are present on the vertebrae. The pelvic girdle may be represented by pubic, ischial, and iliac elements, with traces of pubic and ischial symphyses, or by a single, rodlike bone on each side, or it may be absent entirely. Only the right oviduct is present.

There are two genera in the family. *Typhlops* is widely distributed. It is found in Central and South America, the West Indies, southern Europe, Africa, and southern Asia and it has recently been reported from Australia. *Ramphotyphlops* occurs in the East Indies and Australia.

FIGURE 17-2
Dominican Blind Snakes, *Typhlops dominicana.*

Very little is known of the breeding habits of these small and secretive creatures. The Braminy Blind Snake, *Typhlops braminus,* lays two to seven tiny, elongate eggs, each about 12 mm long and 4 mm in diameter. But apparently not all species in the genus are oviparous, for a specimen of *Typhlops diardi* was found to contain 14 well-developed embryos.

Family Leptotyphlopidae

The Slender Blind Snakes or Thread Snakes are small, degenerate, burrowing forms that bear a close resemblance to the members of the Typhlopidae but differ from them in many structural features. No teeth are present on the upper jaw or roof of the mouth and the maxilla borders the mouth instead of being placed transversely. Rows of teeth appear on the mandible. The pelvis consists of the ilium, ischium, and pubis, but is not attached to the vertebral column. A vestigial femur is present and may project through the skin in the anal region. The cylindrical body is covered with uniform scales; the eyes lack brilles but are covered by the head shields. Only the right oviduct is present. The largest species is only about 300 mm long; the smaller forms are slightly more than 100 mm long.

FIGURE 17-3
The Texas Blind Snake, *Leptotyphlops dulcis.* In the Old World, members of the genus *Leptotyphlops* are often called Thread Snakes.

The species in this family are all included in a single, widely distributed genus, *Leptotyphlops,* which is found in Africa, southwestern Asia, southwestern United States, and tropical America.

As with many subterranean creatures, the life history of the Thread Snakes is poorly known. They live beneath the surface of the ground, but may come out during the early evening to wander for a short time. They feed largely on termites, adroitly sucking the contents from the termite abdomen. There are usually four eggs, which are long and slender.

Family Anomalepidae

Like the other Scolecophidea, members of this family are wormlike burrowers with cylindrical bodies and short tails. They differ from members of both the previous families in having rows of teeth on both jaws and, usually, in the absence of a pelvic girdle. The left oviduct may be well developed, vestigial, or absent. The four genera included in this family, *Anomalepis, Helminthophis, Liotyphlops,* and *Typhlophis,* are confined to Central and South America.

INFRAORDER HENOPHIDEA

The Henophidea comprise a number of conservative types, plus some aberrant, mainly burrowing forms. The optic foramen lies between the frontal and parietal bones. Neural spines are usually present as are vestiges of the pelvic girdle and hind limbs. Five families are included in this infraorder.

Family Boidae

This family includes the largest snakes in the world, the huge Pythons, Boa Constrictors, and Anacondas. (The Boa Constrictor belongs to the genus *Constrictor.* Members of the genus *Boa* are not so large.) Many boids, though, are much smaller than these giants. The Rubber Boa *(Charina)* of the southwestern United States averages about 450 mm long. Boas are typically stout-bodied, short-tailed snakes. The palatomaxillary arch is movably attached to the rest of the skull. Teeth are present on the maxillary, palatine, pterygoid, and dentary bones, and sometimes on the premaxillary. The ventral scales form enlarged, transverse plates; the dorsal scales are small and sometimes iridescent. In all but one subfamily there are vestiges of the pelvis and hind limbs (the latter terminate in clawlike spurs that are usually visible on either side of the vent and are longer in the male than in the female).

FIGURE 17-4
The largest Old World snake, the Reticulated Python, *Python reticulata*. It is exceeded in size only by the Anaconda of South America, which has a maximum recorded length of 1143 cm.

Boids occupy a variety of habitats. Many of the small forms are burrowers in sandy soils. Some are arboreal, with short, more or less prehensile tails. The huge Anaconda of South America *(Eunectes)* is largely aquatic and can remain submerged in water for a long time. A number of the large boids,

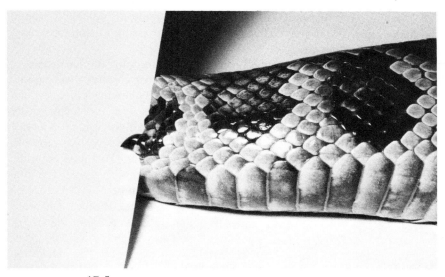

FIGURE 17-5
The external portion of the hind limb of the Cuban Boa, *Epicrates angulifera*.
These spurs apparently are functional during courtship.

FIGURE 17-6
The Cuban Boa, *Epicrates augulifera*, in its characteristic feeding posture.

such as the Indian Python *(Python molurus)*, seem to be equally at home in trees or in water.

Boids feed largely on birds and mammals, and usually kill their prey by constriction. Contrary to popular opinion, they do not crush the bones of their victims. Two or three coils of the snake's body are wrapped around the upper trunk of the prey. These exert enough pressure to stop breathing, and the animal suffocates.

The vestigial hind limbs of the boids are apparently functional structures that are used in courtship. The male Boa Constrictor vibrates them rapidly and rubs them on the back and flanks of his mate. By beating with his spurs, the male Anaconda stimulates the female to move her body into a copulatory position. A male Python, after having arranged himself alongside the female with the anterior part of his body on her back, taps the region around her cloaca with his claws in a slow and rhythmic manner. This may continue for as long as two hours and stops only when the female inclines her anal region on the side and allows the male to insert his hemipenis in her cloaca.

Both oviparity and ovoviviparity are displayed by the Boidae. The Reticulated Python lays eggs, the number of which depend partly on the size of the mother. A large, full-grown female may lay as many as 100 eggs, whereas small females have been known to lay as few as 15. The incubation period lasts 60 to 80 days and the newly hatched young are 600 to 750 mm long. Pythons brood their eggs; the mother coils around them and provides extra heat through an elevation of her body temperature. The amount of such a temperature increase is still a matter of debate. Some authors have recorded the body temperature of the brooding animal to be no higher than that of the surrounding environment, whereas in other species, at other times, body temperatures for the female on the eggs have been reported to be 6° or 7°C warmer than for the nonbrooding male, and 12° to 15°C warmer than the air. From these divergent results, it seems that the body temperature of the female may increase during the brooding process but that the amount of increase varies from one species to another. The temperature of the female is higher at the onset of brooding than toward the end of it. Many more observations of temperature and brooding are needed before the over-all pattern can be determined.

There are also many ovoviviparous snakes in this family. The Anaconda *(Eunectes)* of South America gives birth to between 4 and 39 young (each about 800 mm long) and the Rosy Boa *(Lichanura)* has been known to give birth to 6 young (each about 280 mm long).

The boids are divided into four subfamilies.

Subfamily Boinae.   The great majority of the boas and pythons (about a dozen genera) are placed in this subfamily. It is customary to speak of the New World forms as boas and those of the Old World as pythons, but the two groups are obviously closely related and should not be placed in separate subfamilies simply because they are separated geographically. Usually hypapophyses are present only on the anterior dorsals. The left lung is well

developed in most species, though it is smaller than the right. This is essentially a tropicopolitan group, found in warm countries throughout the world.

Subfamily Bolyeriinae.   Two genera, *Bolyeria* and *Casarea,* found only on Round Island (a small island off Mauritius, near Madagascar) make up this subfamily. They are rather small, semifossorial snakes, differing from all other boids in that the pelvic girdle and hind limb rudiments are completely absent. Hypapophyses are present on all dorsal vertebrae and both lungs are well developed.

Subfamily Erycinae.   This subfamily includes four genera of small boids. *Charina* (Rubber Boa) and *Lichanura* (Rosy Boa) are found in western North America, *Eryx* (Sand Boa) in Africa and Asia, and *Engyrus* (Pacific Boa) in the East Indies. They are short-tailed, small-eyed, mainly fossorial snakes which resemble the Boinae but lack zygapophyses on the caudal vertebrae. Apparently all the Erycinae are ovoviviparous.

Subfamily Loxoceminae.   This small subfamily includes only one very primitive, burrowing species, *Loxocemus bicolor*. It is found in southern Mexico and Central America. Vestiges of the pelvic girdle are present, the left lung is large, posterior hypapophyses are lacking. The scales bear tubercles.

Family Aniliidae

These are stout-bodied, short-tailed, cylindrical snakes which may be as long as 900 mm. The scales are small and smooth, those on the ventral side being slightly enlarged. The bones of the skull are solidly united. Teeth are present on the maxillary, palatine, and pterygoid bones, on the dentary bone of the lower jaw, and, in one genus *(Anilius),* on the premaxillary as well. Hypapophyses and haemopophyses are absent. A vestigial pelvis and rudimentary hind limbs are present, the latter projecting as clawlike spurs on either side of the vent.

The family includes three genera of burrowing snakes: *Anilius,* the beautiful red and black False Coral Snake of northern South America; *Anomochilus* of Sumatra; and *Cylindrophis,* the Pipe Snake or Two-headed Snake of southeastern Asia.

Snakes in which the bones of the skull are solidly fused cannot open their mouths as widely as can most snakes and hence they feed largely upon insects and worms. However, *Cylindrophis rufus* is reported to feed upon other snakes and eels, and to be able to dispose of prey even longer than itself. Aniliidae are ovoviviparous; *Cylindrophis maculatus* produces two or three young that may be more than 125 mm long at birth.

Family Uropeltidae

The Uropeltidae (rough-tails) are secretive, frequently fossorial snakes with rigid, cylindrical bodies and very short tails. Most are less than 600 mm long. Many species are brightly marked with red, orange, or yellow; some are a shiny, iridescent black. The pupil of the eye is round. The ventral scales are little larger than the dorsals. The bones of the skull are more solidly united than are those of any other family of snakes. Burrowing snakes dig with their heads, and the remarkably solid skull of the rough-tails is probably an adaptation to this mode of life. The maxilla has 6 to 8 teeth; the mandible, 8 to 10; and the palatine, 3 or 4 minute ones or none at all. The occipital condyle projects markedly beyond the back of the skull. There are no vestiges of hind limbs or a pelvic girdle.

The most striking characteristic of the rough-tails is the enlarged scale at the end of the tail. It is either very rugose, or spiny, or reduced to two short points. The tail of a freshly caught specimen often is coated with mud. The purpose of this unique appendage has never been satisfactorily explained.

There are seven genera in the family. All are found in damp places in mountainous regions of southern India and Ceylon. These snakes are quiet and inoffensive; they do not bite when handled, nor do they apparently show any fear. When picked up they do not try to escape but entwine themselves around the fingers or a stick and remain in that position for a long time. They are easily kept in captivity and have been known to eat immediately after being caught. Like most small burrowing snakes, they feed on worms and soft-bodied larvae. So far as is known, all are ovoviviparous, producing from three to eight young at a time.

Family Xenopeltidae

This small family contains only one species, *Xenopeltis unicolor,* the Sunbeam Snake of southeastern Asia. Its common name comes from the highly iridescent scales. As it crawls in sunlight it flashes with electric blue, emerald green, blood red, purple, and copper. This brilliant display is seldom seen. however, for these snakes are secretive and largely nocturnal. The body is cylindrical, the tail is short, and the ventral scales are enlarged to form transverse plates. The female may be 900 mm long or more, but the male is somewhat smaller. The bones of the skull are united. The teeth, which are small, are close together and sharply curved; there are four or five teeth on each of the premaxilla and 35 to 45 on each maxilla. The palatine, pterygoid, and dentary also bear teeth. Hypapophyses are absent on the posterior dorsal vertebrae. There is no trace of the pelvic girdle or hind limbs. *Xenopeltis* has two well-developed lungs, the left one being about half as large as the right.

Sunbeam Snakes are frequently found beneath logs and stones in rice fields and gardens near human habitations. They can burrow rapidly in soft earth and those kept in captivity usually spend the day under cover and go forth only at night. They feed on other snakes, small rodents, and frogs. Apparently nothing has been recorded about their breeding habits.

Family Acrochordidae

The weird-looking wart snakes are blunt-headed, small-eyed, ungainly creatures with unusually stout bodies for snakes. The longest females may reach lengths of 180 cm. The skin is loose and the head and body are covered with small, granular or tuberculate, juxtaposed scales. The ventral scales are not enlarged, and those of the head are minute and sometimes pointed in the region of the nostrils; this is the source of the common name, wart snake. (They are also known as elephant's trunk snakes.) Hypapophyses are well developed on all the trunk vertebrae. Pelvic girdle and hind limbs are lacking. The tail is short and compressed.

FIGURE 17-7
Acrochordus javanicus, the Javanese Wart Snake.

Wart snakes are aquatic fish eaters; they are found in estuaries and enter the sea quite freely. Lacking enlarged ventral shields, they are unable to glide normally on land, but progress by a slow, clumsy, heaving of the body. Since they are ovoviviparous, they do not need to come ashore as do egg-laying species. One female has been reported as giving birth to 27 young.

The family includes only two genera, *Acrochordus* and *Chersydrus,* which are found in India, Indo-China, and the Indo-Australian archipelago.

INFRAORDER CAENOPHIDIA

This huge infraorder includes most of the snakes of the world. Its members occupy a wide variety of habitats; some are terrestrial, some arboreal, some fossorial, some aquatic in fresh water, some marine. Paralleling this diversity of habitats is a great diversity of structure so that it is difficult to find characters to define the infraorder. In addition, nothing is known of the internal anatomy of many of the genera. All forms that have been examined so far have only a single carotid artery whereas members of the other infraorders have two. There are never any vestiges of the pelvic girdle. The facial bones are movable and loosely attached to the skull (see Fig. 17–1). The optic foramen is usually bordered by the frontal, parietal, and parasphenoid bones. Members of this infraorder range in sizes from very small to very large, though none approaches the great boas in bulk. Most are harmless to man, but the infraorder also includes all the venomous snakes of the world.

Many attempts have been made to divide this huge and unwieldy assemblage of forms into families and subfamilies. We adopt a conservative position and recognize three families, each with two or more subfamilies.

Family Colubridae

The majority of snakes are included in this family. Usually the belly scales are as wide as the body. Teeth are normally present on the maxillary, palatine, pterygoid, and dentary, but are never found on the premaxillary. Most species have solid teeth, without grooves (aglyphous), and unconnected with any venom glands. A few have several of the rear teeth grooved (opisthoglyphous). The supralabial gland above is specialized to produce a venom which is channeled down the grooves. Such rear-fanged snakes do not inject venom by striking, but by chewing an object that has been taken into the mouth. The venom is thus used, not for capturing prey, but for quieting the struggles of the animal being swallowed, and for initiating digestion. Most rear-fanged snakes are small and harmless, but both the African Boomslang *(Dispholidus)* and the African Vine Snake *(Thelotornis)* have caused human fatalities, and several other species may also be dangerous.

As would be expected in such a large and varied family, colubrids show great diversity in feeding habits. Some will eat almost anything they are able to catch and engulf, whereas others, such as the egg-eating and snail-eating snakes, have specialized diets. The smaller forms eat worms and insects, and many of the larger ones feed exclusively on birds and mammals and usually

FIGURE 17-8
The Blunt-headed Tree Snake, *Imantodes cenchoa*, a Neotropical species.

kill their prey by constriction. Aquatic colubrids prey on fishes and amphibians. The king snakes *(Lampropeltis)* seem to be especially fond of other snakes.

Oviparous, ovoviviparous, and viviparous forms are present in the Colubridae. The life history of the Racer *(Coluber constrictor)*, of the eastern United States, is typical of the oviparous members of the family. Mating has been observed in May, and the eggs are laid in decaying vegetable matter through June and early July. There may be as many as 25 eggs in a clutch, but the average is around 12. Eggs range from 26.8 to 46.5 mm in length, and are somewhat elongated and granular in texture. The young, which hatch in August, are between 200 and 300 mm long.

FIGURE 17-9
Dryophis nasutus, the Indian Green Tree Snake.

The breeding habits of the Northern Water Snake, *Natrix sipedon sipedon,* may be taken as normal for the ovoviviparous forms. Mating takes place in the early spring and the young are born in the late summer and early fall. The average number reported in eight broods was 31, with the number per brood ranging from 16 to 40, but much larger broods have been recorded, some snakes giving birth to more than 75 at one time. Length of the young at birth averages about 225 mm.

True viviparity, with placenta formation, occurs in the Garter Snake, *Thamnophis sirtalis.*

If any generalization can be made about the mode of life history of the Colubridae it is that the more terrestrial forms lay eggs whereas the more aquatic ones give birth to living young. Perhaps this is because eggs laid in the places normally inhabited by aquatic snakes might be endangered by flooding. But even this generalization should not be carried too far, for the highly aquatic Mud Snakes of the genus *Farancia* lay 50 to 100 eggs which the female broods. There is variation even within a single genus: the young of American forms of *Natrix* are born alive, whereas females of the European *Natrix* (less aquatic than their American congeners) lay eggs. On the other hand *Fowlea piscator,* which is as fully aquatic as any North American *Natrix,* lays eggs in late March in the delta of the Indus at the onset of the dry season. The very small young are first to be found when the rains moisten the soil in early July.

Such a large and unwieldy family as the Colubridae can be much more easily discussed if it is subdivided into smaller groups, or subfamilies, but this is very difficult to do. The family is here divided into eleven subfamilies. Eight small groups have been separated from the others. The remainder of this bewildering assemblage has been placed in three large, heterogeneous, rather ill-defined subfamilies. It must be emphasized that this classification is only provisional, and will probably be modified as our knowledge of the anatomy of the colubrid snakes increases.

Subfamily Xenoderminae.   This subfamily includes four small genera of snakes of southeastern Asia and the East Indies—*Xenoderma, Stoliczkaia, Fimbrios,* and *Achalinus.* The neural spines of the vertebrae are expanded and may be flat-topped. Posterior hypapophyses are present. The optic nerve is inclosed between the frontal and parietal. The teeth are aglyphous. The small scales of *Xenoderma* and *Stoliczkaia* are more or less separated from one another, not overlapping as they are in most snakes. The posterior edges of the scales around the lips are upturned in *Achalinus* to the extent that it appears to have a fringe around the mouth.

The habits of these snakes are poorly known. *Achalinus* feeds on earthworms. *Xenoderma* is nocturnal and lives in loose, wet earth. It is frequently found in cultivated fields and eats frogs. Two to four eggs are laid at a time.

Subfamily Pareinae.   The bluntheads are slender little snakes with short, wide heads, slim necks, and big, childlike eyes. Hypapophyses are present only in the cervical region. The teeth are aglyphous. The mouth opening extends far behind the fringe of the buccal membrane. The nasal gland (one of the salivary glands) is enormous. The dentary bone is immovably fused to the surangular. The most striking feature in this subfamily is the arrangement of the scales under the lower jaw. In most snakes these chin shields are separated at the midline by a furrow, the mental groove, which is lined with distensible skin. This allows the two halves of the jaw to be spread widely for the swallowing of large prey. In the Pareinae, the chin shields of the two sides dovetail and there is no mental groove. Consequently, their jaws cannot be spread widely and their diet is restricted to such small items as snails, slugs, and grubs. They apparently cannot crush the shell of a snail, but use their sharp teeth to extract the body before swallowing it. The Pareinae are found in Southeast Asia and the East Indies. There are only two genera— *Aplopeltura* with a single species and *Pareas* with about fifteen species. Some are terrestrial, some arboreal. They are quiet, inoffensive little snakes, mostly nocturnal. So far as is known, they are all oviparous.

Subfamily Dipsadinae.   The slug-eating snakes are the New World counterparts of the Pareinae, which they resemble in appearance and in habits. It is quite possible that they should be placed in the same subfamily. The mental groove is present in some species of slug-eating snakes, but is absent in others. The subfamily ranges from Mexico to Brazil and includes three genera: *Dipsas, Sibon,* and *Sibynomorphus.*

Subfamily Calamariinae.   The Reed Snakes *(Calamaria)* and allied genera are found in the East Indies and Southeast Asia. They are small, secretive, burrowing snakes with smooth, cylindrical bodies, blunt snouts, and short tails. The eyes are small and the number of scales on the head is reduced. Posterior hypapophyses are absent. The teeth are aglyphous. The optic foramen lies entirely between the frontal and parasphenoid bones, a condition not known in any other snakes. The Reed Snakes eat earthworms.

Subfamily Sibynophinae.   These rather small, slender, graceful snakes have well-developed hypapophyses on all dorsal vertebrae and a movable joint in the lower jaw between the dentary and the surangular bones. The numerous teeth are small and uniform in size. There are only two genera, *Sibynophis* of the Orient and *Scaphiodontophis* of Central America. These snakes inhabit the forest floor in mountainous regions, feeding chiefly on lizards. They lay from two to four eggs at a time.

Subfamily Xenodontinae.   Three genera of New World snakes are included in this subfamily. *Heterodon* is found in eastern North America, *Xenodon* in Central America and northern South America, *Lystrophus* in South America. These snakes have enlarged, but ungrooved, posterior maxillary teeth and lack posterior hypapophyses. The adrenal glands are very large. Members of all three genera apparently feed on toads. *Lystrophis* shows threatening and death-feigning behavior that is remarkably similar to that of the Hognose Snakes *(Heterodon).*

Subfamily Dasypeltinae.   This subfamily comprises two genera of egg-eating snakes. For an animal not much thicker than a man's finger to engulf an object the size of a hen's egg is indeed an astounding feat. They accomplish this by a striking series of adaptations. Although there is no mental groove, the skin along the angle of the mouth and cheek region is especially modified for expansion. The teeth are minute, reduced in number, and restricted to the posterior parts of the maxilla, palatine, and dentary, and although the upper jaw elements are rigidly fused together, the bones of the lower jaw are very loosely connected. The most remarkable adaptation of all is the extension of some of the cervical hypapophyses down to pierce the esophagus. The hypapophyses in *Dasypeltis* are highly modified: the ones in front have their ventral edges enlarged into sledlike runners, whereas those behind form elongate, forward-pointing spines. The egg, which is swallowed whole, glides down the runners and is forced against the sharp edges of the spikelike hypapophyses thereby being cut open. The contents of the egg pass into the stomach and the crushed shell is then regurgitated. The hypapophyses that pierce the esophagus in *Elachistodon* are less highly modified in shape; the snakes of this genus have one or two enlarged and grooved teeth on the rear of the maxilla. It is possible that they may also feed on small birds and mammals.

There are about six species of *Dasypeltis* widely distributed in tropical and southern Africa. *Elachistodon* is known only from a few specimens taken in northeastern India. Both genera are apparently oviparous.

Subfamily Aparallactinae.   This subfamily includes several genera of burrowing snakes from Africa and the Middle East. They have short snouts, small eyes, cylindrical bodies, and short tails. Deeply grooved fangs are present in the anterior part of the upper jaw. These fangs are usually preceded, but never followed, by smaller ungrooved teeth; they are thus homologous to the rear fangs of the other opisthoglyphous colubrids rather than to the front fangs of the Elapidae, though members of one genus *(Homorelaps)* were for a long time classified with that family.

Subfamily Homalopsinae.   This large subfamily of opisthoglyphous snakes includes aquatic, terrestrial, and arboreal forms. The aquatic ones are usually stout-bodied and short-tailed, but the arboreal forms are slim and usually have very elongated heads. The Oriental water snakes included in this subfamily *(Enhydris, Herpestes,* etc.) have valves on the nostrils that can be closed when the snake is submerged and narrow ventral scales. The terrestrial and arboreal forms were formerly often included in a subfamily Boiginae. These snakes are largely nocturnal. The Homalopsinae are widely distributed in southeastern Asia and the East Indies, northern Australia, Africa, and tropical America.

Subfamily Natricinae.   The natricines have well-developed hypapophyses throughout the vertebral column, wide ventral scales, and nonvalvular nostrils. Both aquatic and terrestrial forms are included in this heterogeneous assemblage: some are large, others quite small; some are aglyphous, others opisthoglyphous; some are oviparous, others ovoviviparous or viviparous. This large subfamily occurs in Europe, Asia, Africa, North America, and northern Australia. There are many genera and many species, including the familiar water snakes *(Natrix)* and garter snakes *(Thamnophis)* of the United States.

Subfamily Colubrinae.   This is another large cosmopolitan subfamily. The hypapophyses are reduced on the posterior dorsal vertebrae. The nostrils are lateral; the ventral scales are well developed; the teeth may be solid, or the posterior two or three may be grooved. The head is covered with large symmetrical shields.

FIGURE 17-10

The Northern Water Snake of the eastern United States, *Natrix s. sipedon.*
This common snake is called "moccasin" by practically all laymen in its range.

FIGURE 17-11

Natrix erythrogaster, the Red-bellied Water Snake of the eastern United States, in the
act of shedding. The snake will ultimately crawl completely out of the shed skin,
leaving it more or less entire and inside out.

Most of the colubrine snakes are either terrestrial or arboreal. Besides
such familiar types as the racers *(Coluber)* and rat snakes *(Elaphe)* the subfamily includes a number of slim, long-bodied, tree snakes. The famous
"flying snake" *(Chrysopelea)* of the East Indies is an arboreal form that is
able to glide through the air. As it launches itself from a branch it straightens
its body and draws in its belly scales to form a concave surface. Colubrines

FIGURE 17-12
The African Boomslang, *Dispholidus typus,* a snake
whose bite is potentially fatal to man.

vary greatly in size. *Ptyas mucosa,* the Greater Rat Snake of Asia and the
East Indies, may be more than 300 cm long. Most colubrines are oviparous
but a few are ovoviviparous.

Family Elapidae

This is the family of the extremely venomous coral snakes, cobras, mambas,
kraits and sea snakes. Like the members of the following family, Viperidae,
they have venom fangs in the front part of the upper jaw. Snakes with fangs
of this sort are called proteroglyphs. These two families have given all snakes
a bad name, though the venomous snakes comprise but a small part of the
snake fauna of the world. The fang of the Elapidae is a more or less enlarged, canaliculate tooth, which is held permanently in an erect position and
fits into a pocket in the gum tissue on the outside of the mandible but inside
the lip when the jaw is closed. The canaliculate tooth has apparently evolved
from a grooved tooth like those of the opisthoglyphous colubrids. The
groove has sunk in to form a horseshoe-shaped cavity. In the elapids the gap
between the ends of the shoe is usually more or less filled in with calcium
but it still shows as a furrow on the front surface of the tooth. The duct
from the venom gland is not attached directly to the fang but expands into a

small cavity in the gum above the opening of the tooth canal. Two fangs are normally present on each maxilla, lying side by side, though usually only one at a time is firmly attached and functional. Each is followed by a series of developing replacement fangs. Snake teeth are constantly being shed and replaced and this arrangement insures that the snake is never without functional fangs. When the fang on the inner side of the maxilla drops off, the one on the outer side is either ready to be used or is already in use. It serves while the next replacement fang on the inner side is growing into place. The maxillary bone is shortened and probably represents only the hind part of the maxilla found in the rear-fanged, opisthoglyphous snakes. It usually bears one or more small, solid or slightly grooved, teeth behind the fang. Teeth are also present on the pterygoids, palatines, and dentaries. The facial bones are movable. Hypapophyses are developed throughout the vertebral column, and the pelvic girdle and left lung are lacking. There are two subfamilies.

FIGURE 17-13
The defense attitude of the South American Coral Snake, *Micrurus frontalis.*

Subfamily Elapinae. Members of this subfamily are found throughout the tropical and subtropical regions of the world, but they are most numerous in Australia, where most of the snakes belong to this group. They are absent from Europe today but fossil forms have been found from the Miocene and

Pliocene of France. About 30 genera are known, including the longest of all venomous snakes, the King Cobra *(Ophiophagus hannah)* which may be more than 540 cm long.

The snake so often pictured with Indian snake charmers is an Indian Cobra in its defense position. It is not "charmed" but is reacting to the presence of a possible aggressor by a threat display—raising the forepart of its body and drawing up its long anterior ribs to spread the skin of its neck into a hood.

Many of the elapids are unaggressive and seem loath to bite, but their venom is highly toxic. Some of the Cobras of Africa and Asia have the extremely unpleasant trait of "spitting" their venom into the eyes of their enemies. Their fangs are modified to permit the streams of venom to be ejected outward instead of downward. Spitting cobras can spray their poison with great force for up to 180 cm and they seem to show a high degree of accuracy in aiming for the eyes. If the venom is washed away immediately no permanent damage results, but untreated animals become blind.

The cobras *(Naja naja)* of India mate during January and February and their eggs are laid in May. Apparently the pair remains together from the time of mating until the young are hatched; the male may also share in guarding the eggs. Incubation takes between 69 and 84 days. The usual number of eggs is rather low, from one to two dozen, but as many as 45 have been recorded.

Most elapids lay eggs, but the Ringhals or Spitting Cobras *(Hemachatus)* of South Africa and some of the Australian forms are ovoviviparous and *Denisonia* of Australia has been reported to be truly viviparous.

Subfamily Hydrophinae. These are the true sea snakes. The largest reach a length of 275 cm and most species are only about one-third as long, so they hardly qualify as the huge sea serpents of legend, but they may have provided the grounds for many such tales. They differ from the Elapinae mainly in the characters by which they are adapted to life in the sea. The body is more or less laterally compressed posteriorly and the tail is very compressed and paddle-shaped. Except in one genus *(Laticauda)* the nostril opens on the upper side of the snout and can be closed tightly by a valve. The tongue is short so that only the cleft portion can be protruded.

Three genera—*Laticauda, Aepysurus,* and *Emydocephalus*—are less specialized for marine life than are the other sea snakes. Their ventral shields are relatively large, being one-third to one-half as wide as the body. They are able to move well on land and apparently spend a good bit of time there. At least two of the species, and possibly all of them, are oviparous, coming to shore to lay their eggs. They are never found far from land, but live in the shallow coastal waters and river estuaries of southeastern Asia, Australia, and the islands of Oceania. As would be expected, they feed on fishes and are often caught in the nets of fishermen.

The majority of the sea snakes (about 15 genera) are completely aquatic. Their ventral scales are either very small or absent (except those of *Ephalophis*). They are graceful, competent swimmers, though they are seldom found far from shore, seeming to prefer the vicinity of coasts particularly around the mouths of rivers, where the waters are comparatively sheltered. Some species bask on the surface of the water and on days when the sea is calm, they may be seen from the bows of steamers, sometimes by the hundreds. One naturalist reported seeing a mass of snakes forming a line across the surface of the sea about 300 cm wide and nearly 100 km long. The snakes were so closely packed that the line could be seen from several miles. He estimated that the column included millions of snakes—probably the largest congregation ever reported. So far as is known, all of these snakes are ovoviviparous. Although they are so aquatic, at least some species come to shore to bear their young. Female sea snakes of the Philippines have reportedly come onto the smaller islets to bear their young among the rocks and tidal pools. They are found in the Indian and Pacific Oceans, along the Asiatic coast and throughout the Indo-Australian seas to the coast of tropical Australia and the oceanic islands of the southern Pacific. One form (*Pelamis*) has extended its range across the Pacific to the shores of tropical America and westward to Madagascar and Africa.

All of the sea snakes are venomous. The venom of some does not appear to be strongly toxic to humans, but laboratory experiments have shown that the venom of others is even more powerful than that of the cobra. At least four deaths of people in bathing have been attributed to sea-snake bites. A number of nonfatal bites of bathers have also been reported. What induced these snakes to bite is unknown because in general hydrophids can be stimulated to bite only after considerable provocation.

Family Viperidae

These are the snakes in which the whole mechanism for the injection of venom reaches its highest development. The maxillary bone is very short but deep vertically and is movably attached to the prefrontal and ectopterygoid bones. The large fang is on its posterior end but because the bone is so shortened anteroposteriorly the fang lies in the front part of the mouth. The canal for the transmission of venom is usually completely closed so that no external groove is visible (solenoglyphous). At rest, the tooth is folded to lie horizontally along the upper jaw. In striking, the fang is brought forward from the resting position by a movement of the bones forming the palatomaxillary arch, with the maxilla turning like a hinge on the anterior end of the prefrontal. Most elapids, with their smaller fixed fangs, tend to bite and hold on, but the viperids, with their large and powerful fangs, are able to inject a greater amount of venom at the instant of bite and they tend to strike and then draw back. Replacement of the fangs is the same as in the Elapidae, and two fangs are often present at the same time. There are no other maxillary teeth. In other characteristics, the viperids resemble the elapids.

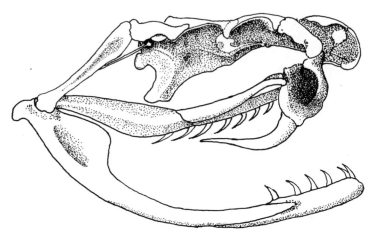

FIGURE 17-14
The skull of the Eastern Diamondback Rattlesnake, *Crotalus adamanteus*.

Snake venoms are highly complex protein mixtures which vary in composition, and hence in effect, from species to species. In general, though, the venom of the elapids acts primarily on the nervous system (neurotoxic) while that of viperids acts on the blood (hemotoxic).

The Viperidae are divided into three subfamilies: the mole vipers (Atractaspidinae), the true vipers (Viperinae), and the pit vipers (Crotalinae).

Subfamily Atractaspidinae.   The members of this subfamily are modified for burrowing; they have a narrow head, small eyes, cylindrical body, and short tail. The fangs are very long and can be made erect with the mouth closed by passing them outside the lower jaw. The posterior hypapophyses are reduced. This subfamily contains only one genus, with about a dozen species in Africa and one in the Sinai Peninsula and Palestine. Apparently they are all oviparous.

Subfamily Viperinae.   The true vipers are usually stockily built, with short bodies and short tails. The large Gaboon Viper *(Bitis gabonica)* may be nearly 180 cm long, but no viperid matches the big cobras in length. The maxillary bone is not hollowed and there is no pit on the side of the face

FIGURE 17-15
An Old World viperid, *Bitis gabonica*, the Gaboon Viper of Africa.

between the nostril and the eye. The nine genera of true vipers are found only in Eurasia and Africa. Some are among the most deadly snakes in the world.

A few of the vipers lay eggs, but in most the young are born alive. A chorio-allantoic placenta forms in the European adder *(Vipera berus)*. Mating takes place in the spring after the snakes have emerged from hibernation. Rivalry between males is keen during the mating season, and they somtimes give a spectacular performance called the "dance of the adders" (see page 117). Although copulation may take place in April, ovulation does not occur until the end of May. The young are born during August and September. The eggs, which will produce the young of the next year, are formed . in the ovaries following parturition. In the northern part of the range, the short summer is apparently insufficient for the complete intraovarian development of the eggs and the snake is forced back into hibernation before the young are ready to be born. In consequence, the vipers in the northern half of Sweden and in Finland breed only every second year. Numbers of young range from 6 to 20, but there are usually about 10 to 14. They may be 130 to 180 mm long.

Subfamily Crotalinae. These are vipers in which the maxillary bone is hollowed out above by a pit that opens between the eye and nostril. The

FIGURE 17-16
A New World pit viper, *Lachesis muta stenophrys*, the Bushmaster of South America.

membrane in this pit is extremely sensitive to changes in temperature and serves to detect the presence of the warm-blooded animals on which the snake preys. Pit vipers occur from eastern Europe across Asia to Japan, and in the Indo-Australian archipelago, but the majority of the species are found in North, Central, and South America. Among the five genera are such dreaded forms as the rattlesnakes *(Crotalus)*, and the tropical American bushmasters *(Lachesis)* and Fer-de-Lance *(Trimeresurus)*. The bushmaster reaches a length of 360 cm and is truly one of the most formidable snakes in the world.

Apparently all of the New World crotalids except *Lachesis* are ovoviviparous but there are some egg laying species of *Agkistrodon* and *Trimeresurus* in southeastern Asia. As in the viperids, there is an extensive premating dance of the males. The Cottonmouth Moccasin *(Agkistrodon piscivorous)* mates in March and the young are born in late August and early September. The number in a brood varies from 5 to 15 with an average of about 8. Length of the young ranges from 150 to more than 250 mm.

READINGS AND REFERENCES

Angel, F. *Vie et Moeurs des Serpentes*. Paris: Payot, 1950.

Bellairs, A. d'A. and G. Underwood. "Origin of Snakes." *Biological Reviews of the Cambridge Philosophical Society*, vol. 26, 1951. (A fairly recent paper that has already become a classic.)

Boulenger, G. A. *Catalogue of the Snakes in the British Museum (Natural History)*. Vols. 1-3, London: British Museum (Natural History), 1893-1896.

Cope, E. D. *The Crocodilians, Lizards, and Snakes of North America*. Report of the United States National Museum for 1898, Washington Smithsonian Institution, 1900.

Klauber, L. M. *Rattlesnakes*. 2 vols. Berkeley and Los Angeles: University of California Press, 1956. (Probably the finest monograph ever published on a single group of herptiles.)

Parker, H. W. *Snakes*. New York: W. W. Norton, 1963.

Pope, C. H. *Snakes Alive*. New York: Viking Press, 1942. (A good, sound introduction to the biology of snakes.)

———. *The Giant Snakes*. New York: Alfred A. Knopf, 1961.

Smith, M. A. *Fauna of British India. Reptilia and Amphibia*. Vol. III. *Serpentes*. London: Taylor and Francis, 1943.

Underwood, G. A. *A Contribution to the Classification of Snakes*. London: British Museum (Natural History), 1967. (The most thorough and comprehensive account ever published on the anatomy of snakes. See also a review of this work by R. Hoffstetter in *Copeia*, vol. 1968, no. 1, 1968. These two publications are the basis of the classification used here.)

Wright, A. H., and A. A. Wright. *Handbook of Snakes*. Vols. 1 and 2. Ithaca, N.Y.: Comstock, 1957. (Contains a wealth of natural history information, though limited geographically to the snakes of the United States and Canada.)

18

RHYNCHOCEPHALIANS AND CROCODILIANS

The two remaining orders of reptiles are both relics. The Tuatara *(Sphenodon punctatus)* is the only surviving member of the order Rhynchocephalia, now placed in the subclass Lepidosauria to which the Squamata also belong. No fossil remains of *Sphenodon* have ever been found, but all other members of the family are known only from the Mesozoic. In spite of this enormous time gap, *Sphenodon* seems to have changed little through the ages and remains today as a relatively unspecialized representative of the reptiles of the early Mesozoic. It has been aptly called a "living fossil." The modern crocodilians are all that remain of the mighty archosaur stock that once throve and gave rise to the Mesozoic dinosaurs, as well as to the modern birds.

Although they are placed in different subclasses, in one respect *Sphenodon* and the crocodilians resemble each other and differ from all other living reptiles. They have diapsid skulls, which have both dorsal and temporal fossae with their bounding arches. The turtles have anapsid skulls, without fossae, and the lizards and snakes have lost one or both of the arches.

ORDER RHYNCHOCEPHALIA

These primitive reptiles look like lizards, but differ from them not only in having a two-arched skull but in many other structural characters. Teeth are present on the premaxillary, maxillary, palatine, and dentary and vestigially on the vomer. Well-developed gastralia are present. The male lacks a copulatory organ. The cloacal opening is a transverse slit. A nictitating membrane, or third eyelid, can be moved slowly across the eyeball from the inner corner of the eye outward while the upper and lower lids remain open. A well-developed parietal eye, with small lens and retina, is present on top of the head. In the young it can be seen clearly through the translucent covering scale, but in the adult the skin above thickens. A similar structure is present in many lizards. It may be sensitive to heat and light.

Family Sphenodontidae

Sphenodon punctatus is scarcely known to most nonzoologists, but to the professional it is one of the most fascinating creatures alive. It shows us, in the flesh, what some of the early reptiles of the Mesozoic must have been like. One of the most perplexing things about it is its apparent failure either to evolve or to become extinct. If we could learn why it has remained virtually unchanged through such a long time, we might understand more of the forces that do bring about evolutionary change in most living things.

The adult Tuatara is about 500 to 800 mm long. It is brownish-olive, and has a small yellow spot in the center of each scale. Enlarged scales form a crest down the back and tail.

Tuataras are found only in New Zealand. Those on the main islands succumbed rapidly to the onslaught of the mammals introduced by European settlers, and the remainder now inhabit only the waterless offshore islands, where they are rigidly protected. They live in close association with vast colonies of nesting shearwaters, called Mutton Birds by the New Zealanders. These birds nest in underground burrows, which they share with the Tuataras. For the most part the association seems amicable, although the normally insectivorous Tuataras occasionally feed on eggs or nestling birds.

The Tuatara remains in its burrow during the day and prowls at night when the temperature drops sharply and cold gusts of wind sweep over the islands. These animals are active at lower temperatures than other reptiles, and their body temperature tends to be lower than that of their surroundings. Body temperatures ranging from 6.2° to 13.3°C have been reported in nature.

Mating has been observed in both the field and the laboratory. Insemination takes place by simple cloacal apposition. During the summer (which, south of the equator, is from November to January) the female lays about ten white, hard-shelled, elongate eggs about 28 mm long. They are usually deposited well away from the home burrow in a shallow hole in sand where they can be warmed by the sun. By August, the embryos are nearly mature. However, the late stage embryo apparently undergoes a sort of aestivation over the second summer and does not hatch until it is about 13 months old. During this aestivation period, the nasal chambers become blocked with a proliferating epithelium that is resorbed shortly before hatching.

ORDER CROCODILIA

Crocodilians are elongated reptiles with a muscular, laterally compressed tail, a more or less elongated snout, and two pairs of short legs. There are five toes on the front feet, and four on the hind feet. A nictitating membrane is present. The cloacal opening is a longitudinal slit, the penis is single, and there is no urinary bladder. The tongue is not protrusible. Gastralia and dorsal and ventral epidermal scales, reinforced by bony plates, are present. The skull is diapsid.

In some respects, crocodilians are the most advanced of all living reptiles. They possess a true cerebral cortex and a completely four-chambered heart. The teeth are thecodont (set in sockets in the jaw). The skin of the head is fused to the skull bones and there are no fleshy lips to make a watertight closure of the mouth possible. The nostrils are far forward on top of the elongate snout. The maxillaries, palatines, and pterygoids meet in the midline of the roof of the mouth to form a secondary palate which separates the nasal passages from the mouth. These air passages open into the throat behind a valve formed by a fleshy fold at the back of the tongue which meets a similar fold on the palate. Water in the mouth is thus kept separate from the inspired air and the crocodile is able to breathe while submerged with only the tip of its snout protruding or while holding prey in the water.

Although *Sphenodon* is of interest mainly to zoologists, crocodilians have a horrible fascination for almost everyone. They truly look like monsters out of the prehistoric past. The huge size attained by some—the maximum length recorded for an American Crocodile is 690 cm—and the man-eating proclivities of a few, so color our concept of the whole group that we think of all crocodiles as ferocious monsters. Actually, some are dwarf forms quite harmless to man. The Congo Dwarf Crocodile has a maximum known length of 112.5 cm.

Evolution in the crocodiles has progressed along two lines. Some forms have a relatively short and broad snout, and the two halves of the lower jaw are joined by a short symphysis. Others have a long, narrow snout, and the symphysis of the lower jaw is long. We recognize eight genera of crocodilians, seven in the family of Crocodylidae and the eighth, *Gavialis*, in a separate family, Gavialidae.

Family Crocodylidae

The snout of the true crocodiles is not sharply set off from the posterior section of the skull. The maxillary bones do not meet dorsally to separate the nasals from the premaxillaries. There are from 14 to 24 teeth on each side of the lower jaw.

The genus *Crocodylus* is widely distributed, being found in the tropical and subtropical regions of North and South America, the West Indies,

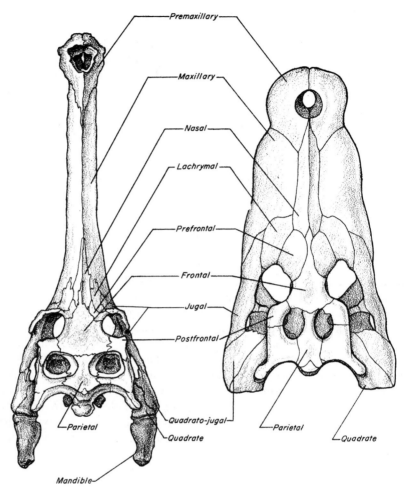

FIGURE 18-1

The skull of *Gavialis gangeticus* (*left*) and of *Crocodylus palustris* (*right*) showing the disposition of the nasal bones (separated from the premaxillaries by the maxillaries in the former but not in the latter). [After M. A. Smith, 1931.]

Africa, Asia, Australia, and the East Indies. It includes both broad-snouted forms, such as the Indian Mugger *(Crocodylus palustris)* and narrow-snouted ones like the African *Crocodylus cataphractus*, in which the mandibular symphysis extends back to the level of the eighth tooth. The genus *Osteo-laemus* comprises two dwarf species found in Africa. *Tomistoma*, the False Gavial of Borneo, Sumatra, and the Malay Peninsula, is the most narrow-snouted of all the Crocodylidae, with the mandibular symphysis extending to the level of the fourteenth or fifteenth tooth.

The broad-snouted, predominantly New World genera are sometimes placed in a separate subfamily, or even family, Alligatoridae, but since *Crocodylus* also includes broad-snouted forms, and since the distinction between the two groups breaks down when fossil forms are considered, it seems best to include them all in the Crocodylidae. *Alligator* is the most northern in distribution of all the crocodilians, being found in the southeastern United States and in southern China. Three genera of alligator-like caimans are found in Central and South America. These are *Caiman, Melanosuchus,* and *Paleosuchus.*

FIGURE 18-2
Young Spectacled Caimans, *Caiman sclerops,* of the Amazon and Orinoco Rivers of South America.

All crocodilians are aquatic, living in rivers, marshes, and lakes, and some take freely to salt water. They frequently bask in the sun on shore but most species seldom wander far from water. The long, strong tail is used for swimming and is also a powerful weapon of offense or defense. By it a victim can be literally swept off his feet and is then seized and dragged into the water before he can recover and escape. If the prey is too large to be swallowed whole the crocodile tears it limb from limb by the gruesomely effective methods of rotating its whole body rapidly over and over in the water.

The sharp pointed teeth are used for seizing but not for chewing. The muscular stomach functions like the gizzard of a bird and, like a bird, a crocodilian swallows hard objects to aid in the grinding of its food. Sixty-nine pebbles were once found in the stomach of a Smooth Fronted Caiman *(Paleosuchus).* In the southeastern United States, where stones are scarce, alligator stomachs have been found to contain Coca-Cola bottles, bottle tops, the brass portions of shot gun shells, and "lightered knots," as the natives call hard knots of resinous pine.

Variation in the number and position of the bony plates that underlie the horny scales in the different species affects the usefulness of the hide. The American Alligator yields a very sturdy and beautiful leather and has been widely hunted. In 1937, 135,000 hides were bought by dealers in Florida at the rate of $4.00 per hide. By 1943 the price had risen to $19.75 for a 7-foot hide, but excessive hunting had so reduced the population that only 6800 were taken. Florida then passed a law protecting the alligator during the breeding season and prohibiting the capture of specimens less than 4 feet long. By 1947 the take had risen to 25,000 hides, and the price had dropped to $13.30 a hide. More recently poaching and excessive drainage of swamps has again reduced the population.

A few lizards and turtles have voices, but for the most part reptiles are silent creatures. Crocodilians are an exception. Young of some species make a curious, high-pitched croak when disturbed. The voice of the adult male is something between a deep bark and a bellow. In the spring months the bellowing of a bull alligator is perhaps the most impressive sound of the southeastern swamps, and is something that every herpetologist looks forward to hearing.

All crocodilians lay large, hard-shelled eggs, either in a shallow excavation on a sandy shore or in a nest piled by the mother. The female American Alligator scoops mud and vegetation with her jaws to build a mound about 90 cm high and 150 to 210 cm wide at the base. She deposits 20 to 70 eggs in a hollow in the center and covers them with material from the rim of the nest. She remains on guard nearby for 9 or 10 weeks until she hears the young, now ready to emerge, beginning to peep loudly. She then tears open the nest and allows them to escape.

Family Gavialidae

The Gavial has an extremely long and slender snout which is sharply set off from the posterior part of the skull. There are 25 or 26 teeth on each side of the lower jaw and the mandibular symphysis extends to the level of the 23rd or 24th tooth. The maxillary bones meet dorsally to separate the nasals from the premaxillaries. The maximum recorded length of the Gavial is 645 cm.

The single living species, *Gavialis gangeticus*, is found only in India and Burma. Its colloquial name in India is "gharial." The generic name was based on this, but through a clerical error was published as *Gavialis* rather than *Garialis*. Now the misformed scientific name has given rise to a new common name, Gavial, which has largely replaced the original Gharial.

The very narrow snout is an adaptation for eating fish, which the Gavial catches by a sudden, sidewise sweep of the head through the water, A broad snout could not be moved so rapidly. The Gavial has been reported to catch birds and such fair-sized mammals as goats and dogs occasionally, but in spite of its large size it seldom if ever attacks man.

The female lays 40 or more eggs in a hole in a sand bank. These eggs are 85 to 90 mm long by 65 to 70 mm wide. The young, which appear in March and April, are about 375 mm long. Obviously they must have been very tightly coiled within the eggs.

READINGS AND REFERENCES

Cope, E. D. *The Crocodilians, Lizards, and Snakes of North America.* Report of the United States National Museum for 1898, pt. 2, 1900.

McIhenny, E. A. *The Alligator's Life History.* Boston: Christopher Publishing House, 1935. (Based on many years of personal experience with these great reptiles.)

Mertens, R. and H. Wermuth. "Die rezenten Schildkröten, Krokodile und Brükenechsen." *Zoologische Jahrbücher,* vol. 63, no. 5, 1955.

Reese, A. M. *The Alligator and its Allies.* New York: Putnam's 1915.

Smith, M. A. *Fauna of British India. Reptilia and Amphibia.* Vol. I. *Loricata, Testudines.* London: Taylor and Francis, 1931.

CLASSIFICATION

The classification of amphibians and reptiles followed in this book is outlined below. All groups are classified to the ordinal level, and those orders or suborders having living representatives, to the level of families. Groups that are entirely extinct are indicated by asterisks.

CLASS AMPHIBIA

*Superorder Ichthyostegalia
*Superorder Temnospondyli
   *Order Rhachitomi
   *Order Trematosauria
   *Order Stereospondyli
Superorder Lepospondyli
   *Order Aistopoda
   *Order Nectridia
   Order Trachystomata
       Family Sirenidae
   *Order Microsauria

Order Gymnophiona (Apoda)
 Family Ichthyophiidae
 Family Caecilidae
 Family Typhlonectidae
 Family Scolecomorphidae
Order Caudata
 *Family Prosirenidae
 Family Hynobiidae
 Family Cryptobranchidae
 Family Ambystomatidae
 Family Salamandridae
 Family Amphiumidae
 Family Plethodontidae
 Family Proteidae
 *Family Batrachosauroididae
 *Family Scapherpetonidae
Superorder Salientia
 *Order Proanura
 Order Anura
  Suborder Amphicoela
   *Family Notobatrachidae
   Family Ascaphidae
  Suborder Aglossa
   Family Pipidae
   *Family Palaeobatrachidae
  Suborder Opisthocoela
   Family Discoglossidae
   Family Rhinophrynidae
   *Family Montsechobatrachidae
  Suborder Anomocoela
   Family Pelobatidae
  Suborder Diplasiocoela
   Family Ranidae
   Family Rhacophoridae
   Family Microhylidae
   Family Phrynomeridae
  Suborder Procoela
   Family Bufonidae
   Family Atelopodidae
   Family Hylidae
   Family Leptodactylidae
   Family Ceratophryidae
   Family Pseudidae
   Family Centrolenidae
*Superorder Anthracosauria
 *Order Schizomeri
 *Order Embolomeri
 *Order Seymouriamorpha
 *Order Diplomeri

CLASS REPTILIA

Subclass Anapsida
　　*Order Cotylosauria
　　Order Testudinata
　　　　*Suborder Amphichelydia
　　　　*Suborder Proganochelydia
　　　　Suborder Pleurodira
　　　　　　Family Pelomedusidae
　　　　　　Family Chelidae
　　　　Suborder Cryptodira
　　　　　　Superfamily Testudinoidea
　　　　　　　　Family Dermatemydidae
　　　　　　　　Family Chelydridae
　　　　　　　　Family Kinosternidae
　　　　　　　　Family Testudinidae
　　　　　　Superfamily Chelonioidea
　　　　　　　　*Family Toxochelyidae
　　　　　　　　*Family Protostegidae
　　　　　　　　*Family Desmatochelyidae
　　　　　　　　Family Cheloniidae
　　　　　　Superfamily Dermochelyoidea
　　　　　　　　Family Dermochelyidae
　　　　　　Superfamily Carettochelyoidea
　　　　　　　　Family Carettochelyidae
　　　　　　Superfamily Trionychoidea
　　　　　　　　Family Trionychidae
Subclass Lepidosauria
　　*Order Eosuchia
　　Order Rhynchocephalia
　　　　Family Sphenodontidae
　　　　*Family Rhynchosauridae
　　　　*Family Sapheosauridae
　　　　*Family Claraziidae
　　　　*Family Pleurosauridae
　　Order Squamata
　　　　Suborder Lacertilia
　　　　　　*Infraorder Eolacertilia
　　　　　　Infraorder Gekkota
　　　　　　　　Family Gekkonidae
　　　　　　　　Family Pygopodidae
　　　　　　　　Family Xantusiidae
　　　　　　Infraorder Iguania
　　　　　　　　Family Iguanidae
　　　　　　　　Family Agamidae
　　　　　　Infraorder Rhiptoglossa
　　　　　　　　Family Chamaeleonidae

*Family Ardeosauridae
Family Scincidae
Family Anelytropsidae
Family Dibamidae
Family Feylinidae
Family Cordylidae
Family Lacertidae
Family Teiidae
Infraorder Anguinomorpha
Superfamily Diploglossa
Family Anguinidae
Family Aniellidae
Family Xenosauridae
Superfamily Platynota
Family Helodermatidae
Family Varanidae
Family Lanthonotidae
*Family Necrosauridae
*Family Aigialosauridae
*Family Mosasauridae
*Family Dolichosauridae
Suborder Amphisbaenia
Family Trogonophidae
Family Amphisbaenidae
Suborder Serpentes
Serpentes *Incertae Sedis*
*Family Lapparentophiidae
*Family Simoliophiidae
Infraorder Scolecophidia
Family Typhlopidae
Family Leptotyphlopidae
Family Anomalepidae
Infraorder Henophidia
*Family Dinilysiidae
*Family Palaeophiidae
*Family Anomalophiidae
*Family Archaeophiidae
Family Boidae
Family Aniliidae
Family Uropeltidae
Family Xenopeltidae
Family Acrochordidae
Infraorder Caenophidia
Family Colubridae
Family Elapidae
Family Viperidae
Subclass Archosauria
*Order Thecodontia

Order Crocodilia
   *Suborder Protosuchia
   *Suborder Mesosuchia
   Suborder Eusuchia
      *Family Stomatosuchidae
      Family Crocodylidae
      Family Gavialidae
   *Suborder Sebacosuchia
   *Suborder Thalattosuchia
  *Order Pterosauria
  *Order Saurischia
  *Order Ornithischia
*Subclass Ichthyopterygia
  *Order Ichthyosauria
*Subclass Euryapsida
  *Order Araeoscelidia
  *Order Sauropterygia
  *Order Placodontia
*Subclass Synapsida
  *Order Pelycosauria
  *Order Therapsida
Reptilia *Incertae Sedis*
  Order Mesosauria

INDEX